# Building Apple Watch Projects

Discover exciting and fun projects by building brilliant applications for the Apple Watch

**Stuart Grimshaw**

BIRMINGHAM - MUMBAI

# Building Apple Watch Projects

First published: February 2016

Production reference: 1250216

Published by Packt Publishing Ltd.
Livery Place
35 Livery Street
Birmingham B3 2PB, UK.

ISBN 978-1-78588-736-9

www.packtpub.com

# Credits

**Author**
Stuart Grimshaw

**Reviewer**
Fito Toledano Carmona

**Commissioning Editor**
Kunal Parikh

**Acquisition Editor**
Ruchita Bhansali

**Content Development Editor**
Mehvash Fatima

**Technical Editor**
Gebin George

**Copy Editor**
Shruti Iyer

**Project Coordinator**
Kinjal Bari

**Proofreader**
Safis Editing

**Indexer**
Tejal Daruwale Soni

**Graphics**
Kirk D'Penha

**Production Coordinator**
Manu Joseph

**Cover Work**
Manu Joseph

# About the Author

**Stuart Grimshaw** has programmed for Apple computers since the days before OS X and has been involved with developing for Apple Watch since its release. Born in the UK and having lived in Germany and the Netherlands, he is currently an iOS developer in Auckland, New Zealand, where he works on some of Australia–New Zealand's largest video and TV delivery apps and heads the research and development of both watchOS and tvOS applications. He is passionate about the potential of the Apple Watch and Apple TV as well as Apple's Swift programming language and is a keen proponent of beach coding.

I'd like to thank Mehvash Fatima for her months of hard work on this book, and her patient answers to a thousand questions.

Thanks also to Ruchita, Fito and Gebin for their hard work and help on the project.

# About the Reviewer

**Fito Toledano Carmona** is a relentless learner. He started coding at the age of 12. He quit medicine to start his own software business at 20, following which he successfully sold his business to start a career at Apple Inc. At 21, he's building a whole new project and a YouTube blog to tell his story.

Stay tuned on Twitter @fito_tc

To my family, for giving me the tools to become who I am today.

To my mentor Francisco, who always believed in me.

To Enrique and Jim, who are helping me put a dent on the universe.

To my friends Adrian, Gabriel, Joan and José Ascanio, Jesús Manuel, Gonzalo, Salo, Alex, and Pedro who always stood up for me no matter what.

To Jack Coyne, Casey Neistat, Phil Toronto, Gary Vaynerchuk, Rafael Nadal, and Steve Jobs for inspiring me.

# www.PacktPub.com

## eBooks, discount offers, and more

Did you know that Packt offers eBook versions of every book published, with PDF and ePub files available? You can upgrade to the eBook version at `www.PacktPub.com` and as a print book customer, you are entitled to a discount on the eBook copy. Get in touch with us at `customercare@packtpub.com` for more details.

At `www.PacktPub.com`, you can also read a collection of free technical articles, sign up for a range of free newsletters and receive exclusive discounts and offers on Packt books and eBooks.

`https://www2.packtpub.com/books/subscription/packtlib`

Do you need instant solutions to your IT questions? PacktLib is Packt's online digital book library. Here, you can search, access, and read Packt's entire library of books.

## Why subscribe?

- Fully searchable across every book published by Packt
- Copy and paste, print, and bookmark content
- On demand and accessible via a web browser

*This book is dedicated to my nearest and dearest, who had to put up with a partner writing a book while in full time employment. Not something I'd like to take on. She did a splendid job.*

*For Jane*

# Table of Contents

# Preface

*Building Apple Watch Projects* provides you with end-to-end guidance on creating a range of apps for the Apple Watch, covering such essential topics as location frameworks, data storage, communication with the iPhone, and animation, to name just a few. It applies the reader's basic Swift knowledge to real-world programming challenges in an easy, step-by-step manner, starting with a simple animated version of the ubiquitous Hello World app, progressing to apps that are Internet connected, location-aware, and fascinating to use, spanning the genres of productivity, games, and lifestyle apps.

The book contains many tips around making the best use of your coding skills, the tools that surround app development, and the many resources and utilities that exist to make your progress as a developer as smooth and enjoyable as it can be.

By the end of this book, you will have taken apps from the earliest conceptual stages right up to the Store submission.

## What this book covers

*Chapter 1*, *Exploring the New Platform*, introduces the reader to the Apple Watch itself and the many design features that set the context in which the rest of the book is presented.

*Chapter 2*, *Hello Watch*, covers the setup of a new Xcode project and the creation of an uncomplicated but attractive take on the Hello World theme, including a little animation eye candy.

*Chapter 3*, *C-Quence – A Memory Game*, presents a minimalist version of a common memory game and covers the planning and design of an app that will involve more than one screen.

*Chapter 4, Expanding on C-Quence,* builds on the preparations of Chapter 3 to complete a colorful and functional working app, which will also be able to communicate with its paired iPhone to gather textual input from the user.

*Chapter 5, On Q – A Productivity App,* makes use of image assets to add a little sophistication to the user interface of a cue-card app, introduces an interactive menu screen, and makes use of the Taptic Engine to provide feedback to the user.

*Chapter 6, Watching the Weather,* adds Internet connectivity into the mix to create a weather app that fetches its live data from the Web and presents it using a table-based interface. This chapter also introduces the Glance screen, making a portion of the data available to the user without launching the app.

*Chapter 7, Plot Buddy – All about Location,* presents a location-aware app, with which the user can store sets of location data, with or without its paired iPhone. We also see the introduction of Swift protocols and custom initialization methods.

*Chapter 8, Images, Animation, and Sound,* adds icons to the app as well as introducing sequential animations and audio/video media playback. We also look at configuring an app's UI almost purely in code, in addition to class extensions and Xcode asset catalogs.

*Chapter 9, Wear It, Test It, Tweak It, Ship It,* is all about the steps to be taken after coding is finished, including installation and testing on a physical device as well as preparation of everything that is necessary for submission to the App Store.

*Chapter 10, This Is Only the Beginning,* takes a look at some more advanced techniques to make your apps stand out from the crowd, introduces some techniques for improving your workflow, and covers a range of topics that any watchOS/iOS developer will want to add to his or her programming skills, including a number of peripheral tools essential to professional work in a team environment, in order to set the stage for the reader's progress beyond the ground covered by this book.

# What you need for this book

To create and build the code presented here, you will need nothing more than Apple's Xcode software package, which you can download for free in the App Store, and a Mac to run it on. A number of other tools are introduced, such as the OS X Terminal app, which are already installed on your Mac.

Testing the code can be done using Xcode's Simulator app, though the reader is encouraged to run the apps on a physical device, which is important when testing production-ready code (and is much more fun).

# Who this book is for

If you have some basic knowledge of programming in Swift and are looking for the best way to get started with Apple Watch development, this book is just the right one for you!

# Conventions

In this book, you will find a number of styles of text that distinguish between different kinds of information. Here are some examples of these styles and an explanation of their meaning.

Code words in text, database table names, dummy URLs, and user input are shown as follows:

"We can include other contexts through the use of the include directive."

A block of code is set as follows:

```
var sequence: [Color] = []
var nextAnswerIndex: Int = 0
```

When we wish to draw your attention to a particular part of a code block, the relevant lines or items are set in bold:

```
func clearGame() {
    sequence = []
    nextAnswerIndex = 0
    }
```

Any command-line input or output is written as follows:

```
# cp /usr/src/asterisk-addons/configs/cdr_mysql.conf.sample
    /etc/asterisk/cdr_mysql.conf
```

**New terms** and **important words** are shown in bold. Words that you see on the screen, in menus or dialog boxes for example, appear in the text like this: "You will be presented with the **Play** button."

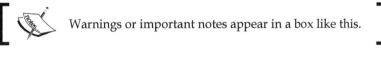

> Warnings or important notes appear in a box like this.

> Tips and tricks appear like this.

# Reader feedback

Feedback from our readers is always welcome. Let us know what you think about this book—what you liked or may have disliked. Reader feedback is important for us to develop titles that you really get the most out of.

To send us general feedback, simply send an e-mail to feedback@packtpub.com, and mention the book title via the subject of your message.

If there is a topic that you have expertise in and you are interested in either writing or contributing to a book, see our author guide on www.packtpub.com/authors.

# Customer support

Now that you are the proud owner of a Packt book, we have a number of things to help you to get the most from your purchase.

## Downloading the example code

You can download the example code files for all Packt books you have purchased from your account at http://www.packtpub.com. If you purchased this book elsewhere, you can visit http://www.packtpub.com/support and register to have the files e-mailed directly to you.

## Downloading the color images of this book

We also provide you with a PDF file that has color images of the screenshots/diagrams used in this book. The color images will help you better understand the changes in the output. You can download this file from https://www.packtpub.com/sites/default/files/downloads/BuildingAppleWatchProjects_ColoredImages.pdf.

# Errata

Although we have taken every care to ensure the accuracy of our content, mistakes do happen. If you find a mistake in one of our books — maybe a mistake in the text or the code — we would be grateful if you would report this to us. By doing so, you can save other readers from frustration and help us improve subsequent versions of this book. If you find any errata, please report them by visiting http://www.packtpub.com/submit-errata, selecting your book, clicking on the **errata submission form** link, and entering the details of your errata. Once your errata are verified, your submission will be accepted and the errata will be uploaded on our website, or added to any list of existing errata, under the Errata section of that title. Any existing errata can be viewed by selecting your title from http://www.packtpub.com/support.

# Piracy

Piracy of copyright material on the Internet is an ongoing problem across all media. At Packt, we take the protection of our copyright and licenses very seriously. If you come across any illegal copies of our works, in any form, on the Internet, please provide us with the location address or website name immediately so that we can pursue a remedy.

Please contact us at copyright@packtpub.com with a link to the suspected pirated material.

We appreciate your help in protecting our authors, and our ability to bring you valuable content.

# Questions

You can contact us at questions@packtpub.com if you are having a problem with any aspect of the book, and we will do our best to address it.

# 1
# Exploring the New Platform

There couldn't have been a more exciting time to be a software developer. A bold claim, undeniably, but if we take a look at the seemingly limitless range of contexts in which we have come to use smart mobile technology for both work and play, it is hard to imagine a more fertile environment in which employ both our creative and technical skills in shaping the next generation of mobile devices.

## Wide open future

It is a rare moment indeed in which developers have the opportunity to use a new programming language, for developing on a new platform, for a new genre of device. Whether relatively new to programming, or with decades of experience across a multitude of platforms and languages, we are, in a sense, all very much beginners, and it is this that many will find the most thrilling part of engaging with the Apple Watch as a developer. We are all in at the ground floor, so to speak, and none of us knows where it will lead us, what users will expect from wearable devices as they become established as mainstream products, what previously unimagined uses will evolve, and what challenges we will face as developers.

As someone who is developing for the Apple Watch, you are truly at the center of this digital revolution. The company that revolutionized our attitudes to computing in general, and mobile devices in particular, is revolutionizing both its hardware and the ways in which developers are able to engage their users, offering them an ever more immersive experience, while the boundaries between hardware and software become increasingly blurred. More than any device before it, the Apple Watch blends into this new landscape, at once a small part in the larger context of mobile computing, and a radical step forward, into a realm of wearable devices that accompany us throughout the day in closest possible proximity.

So before we get down to any coding, we will take a brief look over the concepts, an important undertaking, as we need to understand what users will expect from a device that in some respects resembles others they know already, in order to delight them with things they have never experienced.

In this chapter, we will cover the following

- A closer look at the watch
- A look under the hood
- One app, four interfaces
- User input hardware
- The Watch as extension of the iPhone

# A closer look at the watch

Just as the iPhone presented a revolutionary step in user interface design, so the Watch brings with it a way of thinking about interacting with the user that goes beyond the simple shrinking down of screen content to fit the new device and instead gives us a reimagining of how we both read and input information, while effectively leveraging our previous experience and expectations of using a touchscreen.

The following image shows the Watch's home screen and its older sibling on the iPhone (not to scale):

And as the platform matures there will be endless opportunities, even necessity, for developers to think further out of the box and offer users new experiences and ways to respond to apps that they are just beginning to discover.

# Building on success

The arrival of the Apple Watch took its time, but in 2015 it was time for us to become familiar with a new kind of device, for which the user experience had once again made a complete rethink of the interface necessary. The screen was too small for conventional icons above conventional labels, too small for any kind of menus accessible from a status bar, and generally too small for anything approaching a simple migration of UI elements from the touchscreen of the iPhone. In terms of size, the closest we had was the iPod Nano, but even that 2.5 inch screen allowed a much more conventional interaction with the user than was the case with the Watch.

And once again, the Cupertino design gurus did not disappoint. The home screen, while being recognizably the offspring of the iPhone home Screen, had morphed into a rearrangeable matrix (or **Carousel**) of circular icons that resized as they approached the center of the screen, icons that for the first time had to say all they had to say without the use of a text label, or, it should be noted, without recourse to text within the icon itself — the calendar icon's weekday label really is at the outer edge of readability and should be viewed more as a very neat bit of decoration rather than an attempt to add meaningful textual information to the icon.

A look at the calendar icon on this home screen will, perhaps, make the point better than words:

In addition to the usual gestures of iOS touchscreens, the Watch's touchscreen gained what Apple dubbed the **Force Touch**, adding a layer of input over the familiar one without adding any extra visual elements, the use of which is accompanied by a satisfying tapping sensation produced by the **Taptic Engine**.

Ah yes, the Taptic Engine. Mobile phones have been buzzing away in our pockets and on our table tops for a long time now, with varying degrees of urgency (or sometimes irritation), and other companies' watch offerings have adopted that particular method of notification, as one would expect they would, though with little change in approach. But the Apple Watch—and this really isn't sales hyperbole—takes haptic feedback to a whole new level. The variety, subtlety, and outright aesthetic quality of the Taptic Engine, as Apple has called its tiny stroke of genius, literally has to be experienced to be understood.

So one begins to get an idea of how much we, as Apple Watch developers, have our work cut out for us in terms of rethinking how we will both present information to, and get information from our users, and how their experience in working (and playing) with our apps will be very different to what they, and we, are accustomed to.

# A look under the hood

It is about time we started getting a little technical. We are, after all software developers, and tempting as it may be to look at the many aesthetic qualities of the Watch, we are here to roll up our sleeves and get our hands dirty, and devote ourselves to some technical specifications of the device that will have a direct impact on our work. And the web really has no shortage of images, and articles about, the various and sometimes very luxurious materials and styling that, whether we like to admit it among ourselves or not, are as important to potential early adopting watch owners as all the hard work we do as developers.

So let's start with answering a few questions around the environment in which we will (very soon) be creating our software.

- What hardware are we dealing with, exactly?
- What is watchOS 2?
- What is WatchKit?

We'll start by taking a look some features of the hardware that are relevant to the developer.

# Vital statistics

The Apple Watch comes in two sizes, referred to by the height of the casing, 38mm and 42mm.

 The 38mm model's casing is 33.3mm wide, its height is actually 38.6mm, and houses a display measuring 21.11mm by 26.52mm.

The 42mm model's width is 35.9mm, and has a display measuring 24.34mm by 30.42mm

Both models' casings are 10.5mm deep.

Onto this tiny canvas, Apple have managed to squeeze an impressively sharp and vibrantly colored, pressure sensitive touchscreen, with resolutions of 272 × 340 px on the smaller model, and 312 × 390 px on the larger, giving 290 pixels per inch (PPI) — also often referred to as **dots per inch (DPI)** — enough resolution to satisfy even the most fastidious graphics geek.

The screens of the least expensive model, the Sport model, are, according to Apple:

> *...protected by a lightweight aluminosilicate glass that's especially resistant to scratches and impact.*

Hands up anyone who can tell us about aluminosilicate glass?

All other models are equipped with Sapphire screens; this sounds a little less technical but it is the second hardest transparent material on earth after diamonds. So don't store one in your pocket together with any diamond jewelry.

The 18 hours of battery life claimed by Apple has proved to be a perfectly reasonable estimate, assuming the pattern of engagement that most of us are anticipating at the moment, and even under heavy use and testing during development, the watch is able to go a full working day without a recharge. The user can, when the remaining battery charge becomes perilously low, activate a low power mode that will restrict the watch to telling the time, and your app stays shut out in the cold. We will, of course, later have a look at what power conservation considerations you will want to be including when designing your own apps to avoid this situation.

Apple does the heavy lifting around designing and coding for a single device, one that just happens to come in two sizes, so you will only be working on a single UI design, and only writing one set of source code.

Just to round off the vital statistics, the Watch has 512MB of RAM and 8GB of storage, but the companion app on the iPhone (and there always is one) will store most of your larger resources, images and video, for example, if you have any.

# watchOS

Based on and in many respects similar to iOS, watchOS is the operating system that powers the Watch. It is orientated towards conserving power and the economic use of resources in memory, while offering a rich and engaging experience for the user, and most developers will find leveraging their previous iOS programming experience easy and intuitive, while adapting to the optimizations of many common programming tools and techniques. You will grow accustomed to alternate ways of providing information to the user, for example using tables to display lists, which, while visually similar to those on an iPhone, are handled quite differently at a code level to tables in iOS.

watchOS was revealed to the world along with the Watch in April 2015, and was upgraded to watchOS 2 in the following October. From our perspective as developers, version 2.0 was a huge leap forward, not only as the Watch gained the ability to function away from the iPhone, significantly increasing its scope of use, but also because, just as critically, we gained access to many of the features of watchOS and its hardware that were hitherto reserved for Apple's own apps, most notably the Taptic Engine, Digital Crown, accelerometer, heart rate sensor, speaker, and microphone, and the addition of custom Complications (some of which we will be looking at shortly).

# WatchKit

WatchKit is the Apple framework that will provide you with the technologies that you will use to create your apps. Think of WatchKit as a vast library of code that all developers will need, code that has been written, tested, and optimized for you by the very people that built both the hardware and its operating system, ensuring a maximum level of uniformity, compatibility, code safety, security, and convenience for the developer. There is, to take one example, a massive amount of work for your app involved in redrawing a button when its title changes, work that will be almost identical every time that button changes its appearance in any way, but you as a developer will only write one line of code to get all that boiler plate set up.

This you will likely know already from UIKit, with which iOS apps spring to life.

# One App, four interfaces

Your users have a number of options when engaging with your app, or better said, your app and the user have a number of ways to engage with each other, both parties being able to initiate that engagement:

- The main app
- Glances
- Complications
- Notifications

## The Main app

The main app is the only part of this quartet that is always present in the bundle that gets installed on the watch by the phone app, the other three are optional, and whether to include them or not will depend entirely on what makes sense for your app. This itself adds a new dimension to your app design, as information can be spread across these various channels, partly duplicated, or some combination of both. While the scope and contents of the main app are directly comparable to the equivalent iOS apps, it is worth taking a brief look at what the other three do, how they do it, and what likely use cases they will be associated with.

## Glances

As the name might suggest, the Glance is meant as a brief, and one-way, encounter. The user is presented with a single screen of information, and is offered no interaction other than a single tap, which will take her to the main app. It is an interaction initiated by the user, swiping up from the clock screen. In the images below, you can see that the Glance (left) presents information in a format that is easy to read quickly, containing only essential information, while the main app presents more than just the most immediate data, but requires more than just a casual glance (hence the name) from the user:

Depending on the nature of your app, you may find your users consulting this screen more often than your main app, so it pays to regard the Glance as more than just a stripped down extension of your main app and to give it a level of attention commensurate with its importance. Those who enjoy rising to the challenge of doing a lot with limited resources will likely find this a most pleasing area of design and implementation.

# Correction

It might seem strange to give the name Complication to a UI element, particularly given the emphasis Apple give to the simplicity of using the Watch, but the term actually stems from the tradition of watch-making, and refers to any information, typically dates and weekdays, presented by a watch beyond simply the time of day.

We can see in these images how some watch faces offer different numbers of Complications:

A Complication, as implemented on the Apple Watch, is your chance to present a very small amount of information on the watch face itself, if the user has selected your app from a list of those that offer information in this format. If implemented by yourself and activated by the user, this is also the most direct route to your app, since it involves just a solitary tap on the complication to launch your main app.

# Notifications

This is the one interface that can initiate an exchange between the user and your app. Modelled on and similar in many ways to what we already know from iOS Notifications, your app can draw attention to itself and present to the users a number of options, including launching your main app, or choosing from a set of actions to be undertaken.

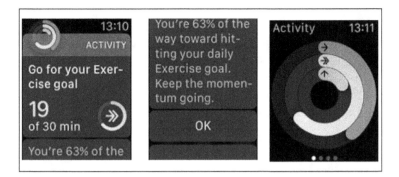

The Notification view (left) can be a scrollable view (center) and may offer the user the chance to choose actions, tapping the app icon launches the main app (right).

Once again, depending on your app's functionality, this may be the interface that your user sees and touches the most. Apple's own Messages app is such a case, in which a user may read all of his messages as they come in, but only react to a limited number of them.

# User Input Hardware

Almost every user will be familiar with swiping and tapping a touch screen as ways of interacting with smartphones and tablets, and the Apple Watch inherits those features in full, but as we have seen already, beginning with watchOS 2 the developer now has access to a number of hardware features built into the device. This is made all the more exciting by the nature of some of these new features, which don't really have any precedent among the devices that preceded the Watch. They include:

- Digital Crown
- Force Touch
- Taptic Engine
- Audio In Out
- Accelerometer

We will not use all of the Watch's hardware features in this book, but in designing an app it makes sense for us to be aware of all the tools at our disposal.

The reader is encouraged to read through, at least once, the Apple Watch Programming Guide, to gain a basic overview of the options available for engaging in a meaningful, original, stylish, and fun way with the user.

`https://developer.apple.com/library/ios/documentation/`
`General/Conceptual/WatchKitProgrammingGuide/`

Much of this stuff really is too good to miss.

# Digital Crown

Possibly more thanks to its snappy name than anything else, one of the most widely reported hardware features of the Watch has been the Digital Crown, which functions both as a push button, comparable in use to the iPhone's *Home* button, and as a method for selecting or entering data into your app. While not exactly a radical development, the royal scroll wheel really does make entering data, selecting options, and navigating around an interface very much quicker and more precise than tapping and swiping what are frequently extremely small UI elements on the screen and also avoids covering a large part of the screen content while doing so; this is the inevitable result of using finger touch gestures.

And if the battery gets a bit low, you could try using it to wind up the watch. Just kidding.

Arousing much less interest than the Crown, and slightly more conventional, is the humble Side Button located below it, which is, well, a button. Everyone needs a button, right? Perhaps to bring up the Friends Screen, with which to send messages and initiate phone calls. It's a fine button.

Oh, and your app can't use it.

# Force Touch

In addition to the standard taps (or touches), and double taps, with which to select options, activate buttons, and so on, in much the same way as she would on any other iOS touchscreen, the user can also deploy the Force Touch, a harder and slightly longer press on the screen, which is acknowledged by watchOS with a satisfying pulse (see Taptic Haptic, below), and can be interpreted by an app in whichever way is appropriate.

# Taptic Engine

Wikipedia offers the definition of haptic technology as

> *...technology which recreates the sense of touch by applying forces, or motions to the user*

Haptic feedback has long been a part of gaming controllers and other devices, including smartphones, made by many manufacturers. Apple's take on this, the Taptic Engine, is definitely one of the industry's classiest implementations of this technology, and employs a linear actuator (look it up, if you like) to produce a quite stunning range of buzzing and tapping sensations on the wrist, that users will have available to them all day, no matter what they are wearing (dresses and smartphones, no pockets, now what?), and no matter where they are, making the Watch into a device that can communicate meaningfully with the user with minimal distraction, maximum discretion (you can't hear the Taptic Engine next to you at all), and one that can initiate that communication (beyond just beeping) in a way that is simply not possible on a phone, which may be on a table in front of the user, in a jacket pocket on a coat hook somewhere, or on a table somewhere quite different.

WatchKit provides a number of preset taptics, which the developer is encouraged to use according to the conventions suggested by the naming of each preset, so that users will, over time, become accustomed to each haptic notification and know intuitively and without needing to look at the watch, what kind of information the watch is offering to convey.

Whether a subtle and gentle tap on the wrist, or an insistent, repeated buzzing sensation, as inaudible as it is invisible, the Taptic Engine represents not only state of the art haptics, but also points the way to a whole new genre of communication from a device to the user. To a great extent, what becomes of this is down to what designers and developers can dream up.

# Audio in out

While the audio output of the Watch is limited to a tiny but impressively loud and clear speaker, one which is loud enough to conduct a telephone conversation in an averagely loud city street (so no headphone output), the audio input of the device is both surprisingly high quality (call recipients are unaware that a conversation is being had from the watch) and pleasingly flexible, in that it gets piped into Apple's voice-controlled assistant **Siri** (yes she lives on the Watch too) where it can be used to dictate text and save audio files, among many other things.

Alternatively, the user can connect to a Bluetooth headset, which is a more suitable arrangement for that jog around the park.

## ♥ It

The **Heartrate Sensor**, in enabling users to monitor one of the most important aspects of their physical health, is undoubtedly an important feature of the Apple Watch. The ability to share heartbeats with another Watch owner perhaps less so.

# Extension of the iPhone, but more

To some extent, users are, by complementing an iPhone with an Apple Watch, extending the reach of the device they already know, and widening the range of ways they have to interact with much of the software they already use. Indeed, at the time of writing, Apple does not accept submissions of apps that provide functionality exclusively for the Watch, demonstrating quite clearly their current view of how the Watch fits into the ecosystem. In this model, the watch is very much an extra level of engagement with the phone, and introduces what may well become a dominant feature of wearables development, namely that the physical separation of two devices does not need to be a fundamental part of the relationship between them. In a sense, it is like taking a small bit of screen and a button or two from the watch, and locating it somewhere more convenient, for use in appropriate situations. Apple's emphasis on fitness tracking is a great example, but should be viewed more as some smart reading of the nascent market for wrist-borne computers, than as an indication of some imagined limit on the scope of the Watch's utility. These are very early days, and it would be a foolhardy author who risks a prediction of the ways in which our use of wearables will evolve in the next few years, that is still very much in the stars. But evolve it will, of that we can be sure.

# Opportunity knocks

And it is because of this wide open scope for development and yet-to-be-discovered potential that we also need to be able to view the Apple Watch as not simply an extension of the iPhone (though it certainly is that), but also as an autonomous device in its own right, one that can collect data from a number of sources (whether the user, the phone or the web), process that data natively, present it visually or otherwise, at its own initiative, and even store that data onboard or online. Since the release of watchOS 2, the limitations of what we can do independently of the phone are defined more by our imagination than by actual barriers inherent in the technology, and the Watch fits in perfectly with the tendency of apps to be increasingly spread across devices and platforms. Most users of social networks, for example, will regularly use more than one platform to catch up on what's going on and update their status, typically a web browser on a laptop and a phone app, but this trend is being broadened across a whole range of software categories, as users increasingly expect to be able to use apps across several devices according to whatever is appropriate to their situation.

And so we have an operating system, watchOS, that is both a part of its larger sibling, iOS, and a separate OS in its own right, that perfectly mirrors the relationship between the physical watch, and its larger counterpart in our pockets. The relationship between WatchKit and UIKit is the same. Our success in designing and developing for this platform will likely depend to a significant extent on our ability to embrace this dual personality.

If that sounds like an exciting challenge, and a stimulating environment in which to expand your skills as a developer, then this book is for you.

# Summary

In this chapter, we have gained an overview of the Apple Watch — the physical device itself, the operating system that powers it, and the wider context in which you will be creating WatchKit apps that make the platform come alive. We have covered details of the hardware, the context in which the Watch idiom is being developed, watchOS, Apple's framework for developing for watchOS (WatchKit), the various interfaces available to your app, the user interface hardware, and some UI objects in Interface Builder

In the next chapter, we will create an app that will run on the Watch Simulator (which is a part of Xcode).

You are, as designer and developer, and as such, pioneer of the new platform, in a unique position to contribute to its growth and success, leveraging its almost limitless potential to transform how we engage with the technology we carry with us (almost) everywhere we go. So let's do it.

# 2
# Hello Watch

Time to fire up the engine and get a feel for some of the new tools we have at our disposal, courtesy of Apple's Xcode development environment. Once we get started, you will recognize much of what you already know through using Xcode and Swift to build apps for iOS, whether for iPhone, iPad, or universal apps for both. The project structure will be familiar to you, and everything that you would expect to see in a purely iOS project will be present.

If you have so far been developing for OS X and not iOS, you will notice many similarities but also many differences to an OS X project. Not a problem. You know Swift already, we will be setting up the projects step by step, and we are all new to WatchKit.

Briefly, we will learn how to do the following:

- Start and set up a project using an Xcode template
- Create some UI elements
- Get the app running on the Watch Simulator
- Add some simple animation to the UI
- Get basic input from the user

By the end of this chapter you will be looking at a simple but functioning WatchKit app, installed and running on the Watch simulator.

# Setting it all up

This will be largely familiar territory to anyone who has already created projects for iOS, with just a couple of things to bear in mind, which will be pointed out as we go along. Do note, however, that if you have not yet updated to Xcode 7 and there are many professional environments in which updating isn't done without a significant degree of caution— then now is the time to do it, Xcode 6 will not build WatchKit apps beyond the first version of watchOS and we will be developing for the vastly superior watchOS 2.

# Creating a new Xcode project

Assuming you have Xcode 7 up and running, we will now let it do much of the work involved in setting up a project that will include the targets, schemes and source files needed to create the basics of a Watch app and its companion iPhone app.

iPhone, by the way, not iPad. Only the iPhone (5 and above) can pair with the Watch, and if the app is run on an iPad, the presence of the Watch app will simply be ignored.

A template is provided by Xcode that will do all this, and though we will look later at other ways of adding a WatchKit target, we will stick with the easiest set up at first.

1. So from the **File** menu select **New | Project...**

2. Select the **Application** template from the **watchOS** section.

Note that we now have the new entry, watchOS, in the list of available platforms; be sure not to choose Apple Watch from the top section, iOS, which is for creating apps for the original watchOS version.

3. Click **Next**.

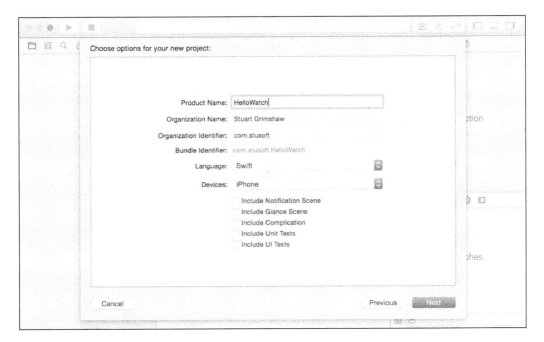

4. Give the project the name **HelloWatch**, and select **Swift** from the **Language** list.

5. From the **Devices** list, select **iPhone**.

6. Deselect all other options as shown.

This is to create as simple and uncluttered a project as possible, and is not intended to discourage the use of unit and UI tests. Similarly, we will be adding other scenes (views) in later projects, but we don't need them at this stage.

7. Click **Next**.

Now save the project file to wherever is appropriate for you. If you usually use version control, there is no reason to do otherwise here, and if you don't, don't worry about it, you can leave it unselected. And there is no need to add the project to another workspace.

8. Hit **Create** and you're done! You now have a fully functioning—though not very functional—Apple Watch app project set up and ready to be turned into something a little more, well, interesting.

# Check out what's new

But before we do that, let's have a look around the project and see what we have set up thus far. Your project window should look very similar to the image here:

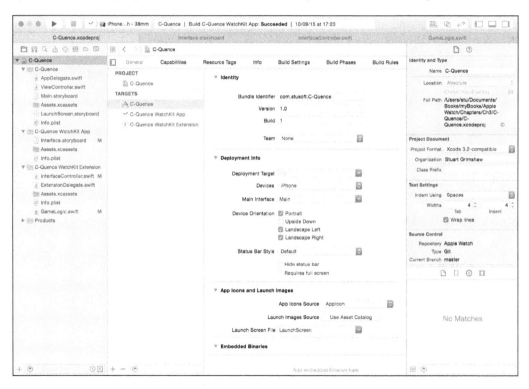

# Building and running the app

If you hit the run button (*Command-R*) now, you'll probably be pleased to note that the app compiles, builds, and runs without complaint. You'll probably also be a little underwhelmed at the result, namely, a blank app window will open in iOS Simulator, which you will have seen before many times no doubt. And that's it. So where's the watch?

If you look carefully at the screenshot below, you will see that the selected *Scheme* is running the iPhone app target. One of the many things that Xcode saved you having to do yourself was creating a scheme for the WatchKit app target, but it is not yet selected.

Select the WatchKit app scheme, and choose one of the phone/watch combinations. I tend to use the smaller watch, the logic being if it's readable on the small one, the larger one will generally present no extra challenges. But whatever suits is fine.

When you run this scheme, one of two things may happen:

If the watchOS Simulator launches, and after showing a watch face for a second or so presents a black screen, with the time in the upper right hand corner, you're in luck. The WatchKit app has been installed and is running, as shown below:

If the clock face is a different one, don't worry, it doesn't matter which is showing. You can even change it, if you like.

However, if you are taken immediately back to the watch face, don't panic. At the time of writing, the watchOS Simulator sometimes needs a second run at it. Just hit **Run** again, and all will be as it should be, and you'll be looking at the image above right.

From now on, running from this scheme will launch the iOS Simulator (and, if necessary, install the iPhone app) as well as the Watch Simulator.

Two Simulators?

You may have noticed that you now have two Simulator apps running (check out the dock). Now that iOS Simulator has to deal with both iOS and watchOS, you will see that it has been divided into just plain 'Simulator' and 'Simulator (Watch)', each part running as a separate app with a separate icon in the dock.

However, to preserve clarity, we will refer to the iOS Simulator and Watch Simulator (Apple have unfortunately decided that such clarity is unnecessary when it comes to the dock icons of the two apps).

iOS Simulator will not come to the foreground, so to see it you'll need to select it from the dock. You may be surprised to see that the phone app is not running; you are simply looking at the home screen. This is because although iOS Simulator is necessary to run the watch app in watchOS Simulator, it is not necessarily the case that the companion phone app needs to be running.

We will look at this again later, for the moment we don't need to deal directly with iOS Simulator.

# Looking over the project

We'll take a moment to have a look at the project, which will differ in some ways to projects you have created just for the iPhone.

# Three apps?

The first thing to notice is that you have more source code files than you are accustomed to seeing when you first create a purely iOS app. These are rather conveniently organized into groups in the Navigator pane as follows:

# iOS app

In the group named **HelloWatch** (assuming you used that name when creating the project) you will find the list of source code, xcasset, and info.plist files with which you are already familiar, and these files all make up the iOS app that will run on the iPhone. Remember, watchKit apps for watchOS need a companion iPhone app, however much the Watch app may be your focus, and possibly your user's.

There is little to add at this stage, and in this chapter we will not be changing any of the files in this group.

# WatchKit app

The group named **HelloWatch WatchKit App** contains the **Storyboard** (and an empty xcassets file) that you will use on the Watch itself. Notice too, the info.plist file, which performs the same function as that of the iOS app, though we won't be editing it yet.

Strangely, perhaps, the WatchKit App group is entirely devoid of any source code. To see anything that looks like Swift, we need to look at the third group created by the template we chose when setting up the project.

# WatchKit extension

In the **WatchKit Extension** group, we finally get to see some source code that will actually run on the watch, and it is in this group that you will want to save the custom classes that you will be creating throughout this book. Now that you have, in effect, three apps to maintain, keeping your code well organized in this fashion has become even more important than it was already. But then, you were totally organized anyway, right?

In addition to source code, this group contains further xcasset resources, and again, a separate info.plist file.

It is the WatchKit Extension that communicates with the iOS app. This is where you will soon be writing code for the Watch.

Why the separation?

In the original incarnation of watchOS, the WatchKit Extension resided not on the watch itself but on the phone, and from there it communicated with the watch (in a sometimes infuriatingly limited and sluggish manner). With watchOS 2, the extension was moved onto the watch itself, where it could run WatchKit code natively, and communication between the phone and watch was completely overhauled, thus enabling a more responsive UI, faster app launches, and improved battery life by reducing the amount of data transfer between the two devices—just one of a multitude of improvements to the platform as a whole.

Keeping this three-way structure meant that it was easy to migrate watchOS 1 apps to the new version, so the trio of apps has, for the time being at least, been retained.

# Three targets

Analogous to the three apps, we now have three **Targets**, each of which has its own build settings (if you have been looking through the source files you might need to select the top level project file in the navigator to see them). You might like to click through the various tabs of each target and compare them, but if any of it seems a little overwhelming, rest assured that we won't be doing much work here for a while yet. Do, however, bear in mind that these multiple targets mean that running and debugging the watchOS and iOS code necessitates the use of the separate schemes we have used already to run the app(s).

# Two storyboards

Two gadgets, two Storyboards. The Watch app and Phone Storyboards are entirely separate.

# Adding some content on screen

So let's get some content onto the screen. We have covered a lot of ground already; negotiated a couple of tricky hurdles and gained a good overview of the project. It really is time to create something more interesting to look at in the Watch Simulator than the less-than-inspiring black screen that shows the time in one corner.

# Preparing the interface

Follow the given instructions to prepare the interface:

1. Open a new tab in Xcode (*Command-T*). In the project Navigator, select the `Interface.storyboard` file from the **HelloWatch WatchKit App** group (make sure you don't select the `Main.storyboard` from the **HelloWatch** group).

2. From the **Editor** menu, select **Show Document Outline** if it is not selected already.

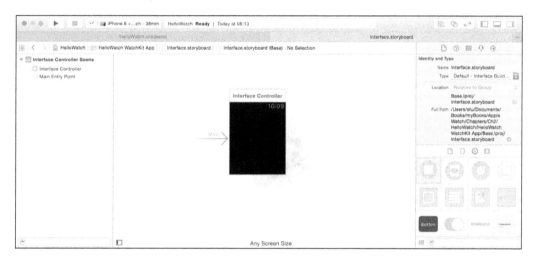

3. Open the Interface Controller Scene in the Document Outline as shown, which will help you navigate around the interface as you add UI elements to it (selecting some of the smaller elements directly can get a little tricky and the outline will help you here).

>  If you don't see the Objects menu in the bottom right hand corner, as shown, hit *Command-Option-Control-3*.

4. As we drag UI elements from the Object menu into the Interface Controller view, the changes will be reflected in the Document Outline.

An arrow labelled `Main` points to the interface, indicating that it is the app's initial controller, and thus the point at which your custom code takes over from the (many) processes that need completing to get your UI onto the watch's screen. When there are multiple interface controllers, this arrow can be moved to whichever controller should load first after app launch. If only it were this simple in iOS.

# Adding a button

To add a button, please follow these steps:

1.  Drag a Button object from the Objects menu onto the interface. Select the Attributes Inspector (*Command-Option-4*) and change the button **Title** to **Hello**.

2.  Set the **Alignment** attributes, both **Horizontal** and **Vertical** to **Center**.

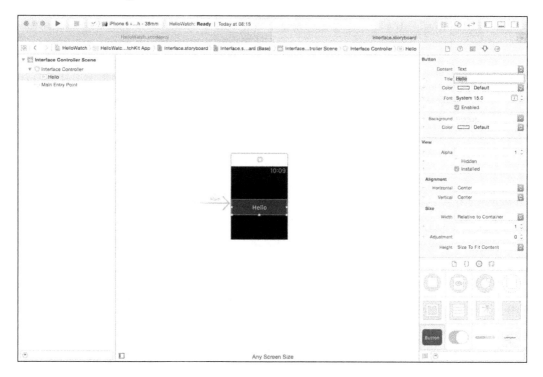

3.  If you hit **Run** now, you'll see the app now has a functioning button, but let's hook it up to a little code first, so that we get at least a small visual result of having tapped it.

4.  Select `InterfaceController.swift` in the project Navigator (possibly in a new tab?) and add the following code to the `class InterfaceController` that was created by the template, immediately after the opening curly brace:

    ```
    class InterfaceController: WKInterfaceController {
    ```

    **@IBOutlet var helloButton: WKInterfaceButton!**

5.  Delete the functions `willActivate()` and `didDeactivate()`, as we will not be using them here.

6.  Add the following function to the code after the `awakeWithContext` function:

```
@IBAction func helloButtonTapped() {
        helloButton.setTitle("World")
    }
```

Now we have an instance variable, `helloButton`, by which we can refer to the button, and a method, `helloButtonTapped()`, that it will call. Your class should now look like this:

```
class InterfaceController: WKInterfaceController {

    @IBOutlet var helloButton: WKInterfaceButton!

    override func awakeWithContext(context: AnyObject?) {
        super.awakeWithContext(context)
    }

    @IBAction func helloButtonTapped() {
        helloButton.setTitle("World")
    }
}
```

However, we still need to connect the button to the relevant sections of code.

1.  Open the assistant editor (*Command-Option-return*) to display the source code alongside the interface, *Control-click* on the `helloButton` you have just created in order to bring up the connections panel, and drag from the **New Referencing Outlet** circle to the `helloButton` variable in the source code:

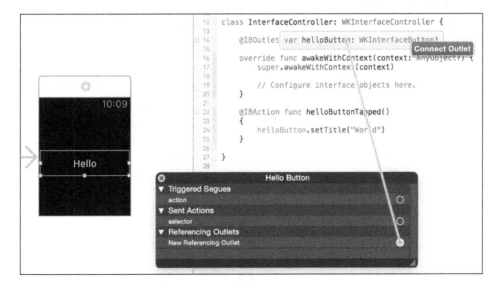

2. Do the same to connect a **selector** from the connections panel to the `helloButtonTapped()` function.

As the number of UI elements increases and the interface becomes crowded, you may find that the easiest way to do this is to *Control-drag* from the button in the Document Outline to the source code in the assistant editor.

Now when you run the app, you will see that the title of the button changes when you tap it; from `Hello` it turns to `World`.

Congratulations! You have designed, coded, compiled, built and run an app. Everything from here on in is just building upon the foundations you have just laid. In common with every `Hello World` application ever written, you won't find many buyers for this one, but we have covered some important material.

And we're not done yet.

# Give the UI some visual appeal

Our button as it stands is working perfectly, but doesn't exactly cry out to be tapped. The button's default (clear) background color and default button font may identify to an experienced user that it is indeed a button, but let's add a little sparkle to what is still a very monochromatic interface so that everyone new to the Watch - and we can, at the moment, be sure that a large proportion of our users will be just that - will both immediately recognize the button for what it is and be motivated to explore it.

# Adding a group

This is a good opportunity to spend a little time looking at one aspect of watchOS development that is very different to its iOS counterpart — UI layout design.

It is not possible with WatchKit to programmatically create an interface or indeed make many fine adjustments to the layout once it has loaded. All this needs to be done before building the app, using the graphical user interface we have been using up until now to add a button to a blank template.

 This GUI is still referred to as **Interface Builder(IB)**, a throw-back to the days when Xcode was strictly for source code, and Interface Builder was a separate application with which to build the user interface). Apple's documentation still refers to IB and Xcode as if they were distinct apps.

Most of the work in laying out the interface is actually done for you by WatchKit, and all you need to do is set a relatively small number of properties on each element that you add. We have already set the button to be horizontally and vertically centered in its containing view, which may be familiar to you from using UIKit's **Autolayout** for iOS. The use of this kind of layout is very much simpler to use with WatchKit. We will dive into more detail on this subject later in the book, but we will use this feature now, in a restrained manner, to overcome the limitations of the button view.

Just as you have less control over the layout of the screen of the Watch than you do with iOS devices, so you also do not have as much control over the appearance of the UI elements provided by WatchKit. The button you have added is an instance of WKInterfaceButton object, not UIButton object, with which you may be used to working. While it makes your life as a developer significantly simpler, WatchKit may not offer you some features that you would prefer to have available, and you will need to think about whether to change your requirements – and given the habits that you will have developed on much larger screens, it seems likely that you will be doing this pretty frequently - or whether the tools at your disposal offer you a less direct route to achieving the desired result.

We will address such a case here. We have decided our button could look a little more button-like. We have two possibilities that are provided directly by the button – we could change the background color to something a little less similar to the interface background, or we could add an image to the button. But both of those feel somewhat overbearing for a single button on an otherwise empty screen - we are looking for something more subtle. A discrete border would do the job, but WKInterfaceButton object offers no such property, so do we need to find something else?

Well it turns out that one of WatchKit's most modest and unexciting looking elements, WKInterfaceGroup object, can be used to great effect, once mixed with a little imagination.

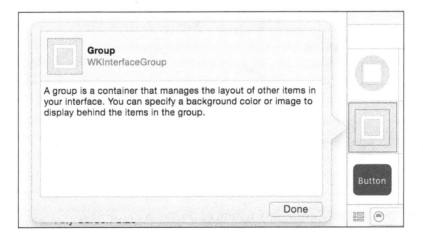

To produce our border around the button, we will simply create a containing group, an instance of WKInterfaceGroup, an object that will contain the button, but be large enough that a small margin is visible around it. And we won't write a single line of code to do it, it's all done in Interface Builder.

# Creating a group object

Follow these steps to create a group object:

1.  Drag a group from the Objects menu onto the interface (it doesn't matter where you drop it) and name it BorderGroup by selecting it in the Document Outline, hitting *Return*, and typing in the name. Hit *Return* again to enter the new name.

2.  In the attributes Inspector, set **Insets** to **Custom** and set **Top**, **Bottom**, **Left** and **Right** to the value **1**.

3.  Choose a color for the border — to make things simple a little later on; use the color picker to select one of the colors provided by UIColor's preset color methods. In the image you can see we have chosen yellow, which is the color returned by UIColor.yellowColor() (which we will be using later):

4.  Set the Group's **Radius** to **Custom** and enter a value of **7**.

5. Set the **Alignment** properties, **Horizontal** and **Vertical**, to **Center**.

6. Set both **Size** properties, **Width** and **Height**, to **Size To Fit Content**.

The Group is now ready; all that remains is to put the button inside it. Again, the easiest way to do this is in the Document Outline—just drag the button onto BorderGroup.

You can also do this directly in the interface, but under some circumstances this can get difficult - when you have very small elements or a large number of them, for example - and sometimes even impossible: When an element is set to be **Hidden**, it conveniently disappears from the interface to show how the UI will look when the view is first loaded and so can no longer be selected.

# Tweaking the button

Set the button **Color** to black. Your attributes should look like this:

# Done

Click outside of the interface so that the button is no longer selected, and take a look at what we've got. The button now appears to have a slim border, and immediately draws attention to itself. (Okay, it doesn't have a lot of competition right now, but you get the point.)

Hit **Run**. Looks great. Looks even nicer on the watch. And we can do even better than that.

# Adding some animation magic

We all know, as developers, the pain of finding that an idea that we had thought simple and quick to implement turns out to be much harder than we had imagined. But when programming with WatchKit, you will often find that some quite advanced-sounding technologies are, in fact, extremely easy to use.

`WKInterfaceController` animation methods are just that, and so we can add some flair to our app by changing the color of the button border – which is to say, the `BorderGroup` background color - using a call to `animateWithDuration`.

First, let's add some code to our `InterfaceController` class, and then we will take a look at what we have done.

1.  *Control-drag* from the Group in the Document Outline to the code in the assistant editor, to create an Outlet connection:

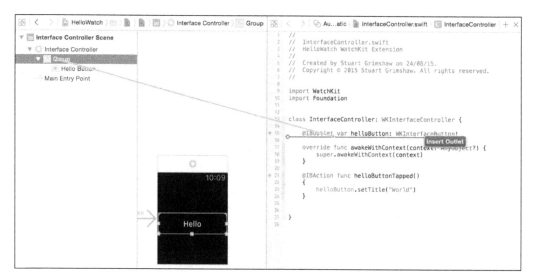

2.  Enter the name `borderGroup` as shown, and hit **Connect**.

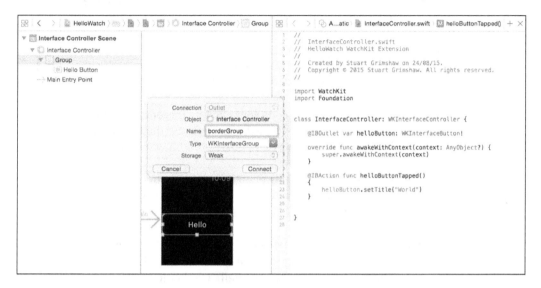

Xcode creates a new `IBOutlet` (connection) between the interface and the source code:

```
@IBOutlet var borderGroup: WKInterfaceGroup!
```

3.  Next, below the `HelloButtonTapped()` method, create a new method with the following code:

```
func changeBorderColor() {
        animateWithDuration(2.5) { () -> Void in
            self.borderGroup.setBackgroundColor(UIColor.
redColor())
        }
    }
```

Why the `self`? Although we won't go into too much detail of the Swift language itself, let's take a brief look at what's happening here.

`animateWithDurations` method takes two arguments, the first is the time that the animation should take (here we have chosen 2.5 seconds) and the second argument is an enclosure that contains the changes in the views properties that we wish to have animated for us by WatchKit.

Because we are sending the code contained by this enclosure to another processing thread, we need it to `capture` the local variables, and for this we must specify where `borderGroup` exists, using the `self` keyword.

If that's a little confusing at this point, it's nice to know that Xcode will throw an error and inform you that you need to insert `self`, should you omit it.

Inside the `animateWithDurations` enclosure, all we need to do is set the `backgroundColor` property of `borderGroup` to a new color, in this case using UIColor's `redColor()` method, which returns, well, a red color.

We call our `changeBorderColor()` function from within `HelloButtonTapped()`:

```
@IBAction func helloButtonTapped() {
        helloButton.setTitle("World")
        changeBorderColor()
    }
```

Your class should now be looking like this:

```
class InterfaceController: WKInterfaceController {

    @IBOutlet var helloButton: WKInterfaceButton!
    @IBOutlet var borderGroup: WKInterfaceGroup!

    override func awakeWithContext(context: AnyObject?) {
        super.awakeWithContext(context)
    }

    @IBAction func helloButtonTapped() {
        helloButton.setTitle("World")
        changeBorderColor()
    }

    func changeBorderColor() {
        animateWithDuration(2.5) { () -> Void in
```

```
            self.borderGroup.setBackgroundColor(UIColor.redColor())
        }
    }
}
```

Go ahead and run the app. When you tap the button, you'll see a slow transition from yellow to red. With just a few lines of code you have added some really sweet looking polish to the interface.

Remember this kind of subtle effect when you are thinking about animations. A search on the Web will show that most blog posts and tutorials tend to concentrate on moving elements on, off, and around the screen, which, as is frequently pointed out, quickly gets tiresome. The use of animations for much more modest hints to the user about the state of the app is a valuable thing to have in your toolkit.

# Getting user input

So far, we have had very little input from the user. She has launched the app and pressed a button, but the Watch is good for more than just presenting information. While the size of the screen makes some forms of input impractical, and text input via a keyboard must be top of that list, there are still many user input methods, both old and new at our disposal.

WatchKit's *text input controller* is a very simple way to gather text input (as the name might suggest) using the presentTextInputControllerWithSuggestions method provided by WKInterfaceController. When making this method call, you provide a list of text options from which the user may make a selection (she may also cancel, or choose voice input).

Firstly, we want to modify changeBorderColor() to accept a String argument which will tell it what color the user has selected. Replace the function as it stands with the following:

```
func changeBorderColor(colorString: String) {
    let newColor: UIColor

    switch colorString {
    case kRed:
        newColor = UIColor.redColor()
    case kPurple:
        newColor = UIColor.purpleColor()
    case kBlue:
        newColor = UIColor.blueColor()
    default:
```

```
            return
        }

        animateWithDuration(2.5, animations: {
            self.borderGroup.setBackgroundColor(newColor)
        })
    }
```

The compiler will complain that it knows nothing of kRed, KPurple, and kBlue. These are constants we will create to prevent any typos creeping into our code, such as red instead of Red. Add these constant declarations directly after the import statements at the top of the code:

```
import WatchKit
import Foundation

let kRed    = "Red"
let kPurple = "Purple"
let kBlue   = "Blue"
```

The compiler warnings will now disappear.

Next, remove the code inside buttonTapped(), replacing it with a call to a new method presentColorOptionsToUser() and add the definition for the new method:

```
func presentColorOptionsToUser() {
    presentTextInputControllerWithSuggestions(
        [kRed, kPurple, kBlue], //1
        allowedInputMode: WKTextInputMode.Plain, //2
        completion:{(results: [AnyObject]?) -> Void in //3
            if let validResults = results, //4
                let resultString = validResults[0] as? String //5
            {
                self.helloButton.setTitle(resultString)
                self.changeBorderColor(resultString)
            }
        })
}
```

This is quite an intense chunk of code, so let us take a detailed look at what we are doing.

- We provide an Array of constant String values that are to be offered to the user.

- allowedInputMode is set to .Plain, since we have no use for emojis in this particular context (!).

- We specify the `completion` closure, which will be called when the user makes a selection (or cancels). This closure takes an optional array of objects, which we have called `results`; this array will contain the selected `String`, or `nil` if no `String` was selected.
- We check that `results` is not nil.
- We check that first object in the `validResults` array is indeed a `String` object and use that the String to set the title of `the helloButton`, and as the argument to `changeBorderColor(colorString: String)`.

Check that your complete code, including the constant declarations, looks like this:

```
import WatchKit
import Foundation

let kRed    = "Red"
let kPurple = "Purple"
let kBlue   = "Blue"

class InterfaceController: WKInterfaceController {

    @IBOutlet var helloButton: WKInterfaceButton!
    @IBOutlet var borderGroup: WKInterfaceGroup!

    override func awakeWithContext(context: AnyObject?) {
        super.awakeWithContext(context)
    }

    @IBAction func helloButtonTapped(){
        presentColorOptionsToUser()
    }

    func presentColorOptionsToUser() {
        presentTextInputControllerWithSuggestions(
            [kRed, kPurple, kBlue],
            allowedInputMode: WKTextInputMode.Plain,
            completion:{(results: [AnyObject]?) -> Void in
                if let validResults = results,
                    let resultString = validResults[0] as? String
                {
                    self.helloButton.setTitle(resultString)
                    self.changeBorderColor(resultString)
                }
        })
    }
```

```
func changeBorderColor(colorString: String) {
    let newColor: UIColor

    switch colorString {
    case kRed:
        newColor = UIColor.redColor()
    case kPurple:
        newColor = UIColor.purpleColor()
    case kBlue:
        newColor = UIColor.blueColor()
    default:
        return
    }

    animateWithDuration(2.5, animations: {
        self.borderGroup.setBackgroundColor(newColor)
    })
  }
}
```

When you run the app, tapping the helloButton will now bring up a modal screen offering you the options seen in the screenshot below:

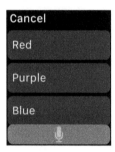

The **Cancel** button is created automatically by WatchKit. Tapping it will dismiss the modal view, and also return nil to the closure that we provided when calling presentTextInputControllerWithSuggestions, and so our results array will be nil. Our code checks for nil, and thus returns without doing anything.

Tapping the microphone icon will also do nothing; it will not even dismiss the modal view, since Watch Simulator does not handle voice input.

As we can see, the list of text options simply mirrors the array you passed as first argument into presentTextInputControllerWithSuggestions. Tapping one of these will dismiss the modal view and return the appropriate String to the closure, and thus we can use the results value to extract that String.

Did you really read all that before tapping one of the options? Either way, on tapping a color, you will see our button's apparent border (really a Group object) animate to its new color and change its title to reflect the selected text.

A subtle but pleasing animation, I am sure you will agree.

# Ideas for self study

Before moving on, it's a good idea to play around with the features and code we have covered in this chapter and explore what happens when changing some of the details to see what effect it has on the app and its interface.

One excellent candidate is the `borderGroup` object we used to fake the button border. If you *option-click* on the declaration at the top of the `InterfaceController` code, you will be presented with a pop-up window that provides you, with among much other useful information, a list of `WKInterfaceGroup` settable properties which are available to you. Start by typing `self.borderGroup.set` and let Xcode's code completion show you what's on offer:

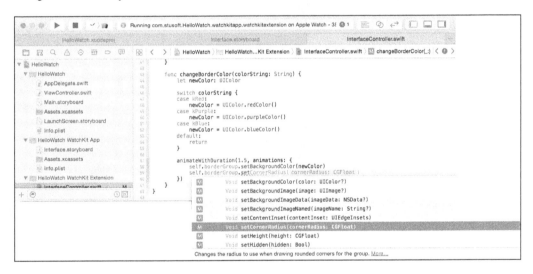

A *Command-click* on one of the UIColor methods such as `.redColor()` will take you to a list of convenience methods which create a limited number of preset colors. Why stick to a choice of three? Why not 10? (But you'll have to scroll!)

You might also like to `print()` a warning to the console when the modal dialog returns `nil` after you select **Cancel** instead of a color from the modal view.

# Summary

In this chapter, you have set up an Xcode project with the necessary targets and schemes to develop the app, and learned to navigate your way around the multiple apps pattern provided by the WatchKit app template. You have added UI elements to the screen, using Interface Builder and linked them to code written in the `InterfaceController` file. One of those elements was the Group object which was used to customize the layout of the UI. You also added animated changes to the UI, and learned how to fetch input from the user using the text input modal view

That is a lot of ground to have covered, so congratulations. You are likely to use most, if not all, of these techniques in just about every app you write, so make sure you have a firm and intuitive grasp of the material we have covered here.

In the next chapter, we will create a simple brain-training game app, for which you will use the power of Group objects to layout other UI elements on screen; add navigation to provide access to other screens; create game logic classes that will reside on the watch; allow the user to further customize the game's appearance; and store user preferences.

# C-Quence – A Memory Game

3

We will be crafting an app that is a little more entertaining by using everything that we covered in *Chapter 1*, *Exploring the New Platform*, adding code which uses basic Swift features that most developers will find familiar and will address some of the topics that face the developer in creating software for a platform that presents some unique challenges.

**C-Quence** will be a game that challenges players' ability to memorize a sequence of colors generated by the app.

It is a game to be played in short bursts rather than prolonged activity, as one of the first things that becomes clear when using a physical device is that the watch is unsuited to tasks that take more than a short time to complete, which we will keep in mind as we look at the top-level design of the app.

Bear in mind that, although this is a very modest app in terms of the amount of coding it takes to bring it to completion, we still want to adhere to what some refer to as Best Practice (and others prefer to think of as simply learning from others' mistakes without the *schadenfreude*).

Here is a brief overview of how we will approach the various steps of development:

- Plan the flow of the app
- Set up the Xcode project
- Build the interface in Interface Builder
- Create the game logic class
- Implement the Interface Controller logic
- Hook up the classes and interface

The code presented in this chapter will reside fully on the watch, needing no support from the iPhone companion app. It will not be a complete and functioning app, that will come in the following chapter, but it will be a robust framework and we will have learned some important principles of software design.

# Plan the app

Let's have a look at how the app will be developed.

# Mission statement

Let us look at the app's requirements before beginning any work in Xcode.

We will write a game that presents a group of four color-coded buttons which flash in an order generated by the app, which is then to be repeated by the user by tapping the buttons. If the user does this successfully, the sequence length is extended by adding a randomly generated color, and is flashed again (the flash interval needs to be a suitable time, say, in the region of one second). The sequence gets longer and longer, each time requiring the user to repeat it. Eventually the user will fail, and will be presented with his 'score' – the length of the last correctly guessed sequence, and a prompt to start the next game.

So that's our game, time to get building that interface, right?

Well, no, not quite yet. To make the best use of our coding time, and to produce the best code possible, we will try to formalize our game requirements into a form that will help us maintain a clear picture of what we are aiming to achieve. Our mission statement (a term to be taken with a pinch of salt, perhaps, but it does sum it up rather nicely) is already the first step in this process. Next we want to take a blow-by-blow look at how the user progresses through the app.

# User story

It's not just film directors and producers that get to play with storyboards; long before Apple commandeered the term for their IB interface layouts, software engineers came together to run through the so-called user stories and sketch the storyboards that emerged from these sessions.

As tempting as it is to start work on a smart looking bunch of buttons and labels that we can see on our simulators and devices (and garner a few oohs and ahs from friends and colleagues), it really is essential to get some kind of handle on what the software will be doing when it comes up against its biggest challenge—the user. Doing this will substantially improve the quality of code that we write and the speed with which we write it. It is also likely to reduce the amount of code we produce and as any experienced developer will tell you, less code means fewer mistakes, easier code maintenance, and greater efficiency.

```
lessCode == betterCode //evaluates to true
```

As the first attempt, let's just run through what the user will experience when using the app, without (yet) attempting to extract any formal specifications.

But this is a game and games don't have users, they have players, so our user story becomes a player story:

The player will launch the app and be presented with a screen prompting her to begin play. This suggests a button (no prizes for working that one out), possibly labelled 'Play'. Whatever we label it, we shall refer to it as the **Play** button. She will have no choice than to tap the **Play** button (or leave the app, but let's be optimistic here) and so we can move onto the next step. The app will show the four colored buttons and flash a sequence of colors, which is generated on the fly by adding a random color to that sequence (the game starts with an empty sequence, so extending the sequence needs to take place before the first round). At this point, we do not want the player to be able to actively engage with the app and so we disable user interaction on the buttons.

Now that we are ready to accept the player's attempt to repeat the sequence, we need to enable user interaction.

The player's guess is evaluated. If the player's guess is correct, but she has not yet guessed the complete sequence, we simply wait for the next guess. If the player's guess is correct and she has guessed the whole sequence, user interaction is disabled once more, the sequence is extended by a random color, and flashed on the screen. If the guess is wrong, the game is over, and we present the player with both her score and the **Play** button with which she has the opportunity to play the next round. Also, we need to reset the sequence back to empty in preparation for the next game.

Now we can arrange that user story into pseudocode, a rough description of what we need the app to do:

Present Play button

- User hits Play

Play sequence

- Extend sequence by one color
- Disable user interaction
- Flash colors in sequence

Get player input

- Enable user interaction
- Store user input
- Acknowledge user input with color flash

If player guesses correctly and guess is incomplete

- Get player input (as above)

Otherwise (i.e. user guesses correctly and guess is complete)

- Play sequence (as above)

If user guesses wrong

- Show results
- Reset sequence
- Present **Play** button (as above)

We might want to sketch out such `pseudocode` in a more visual manner to produce a flow-chart that will provide a map for us to follow when designing the app:

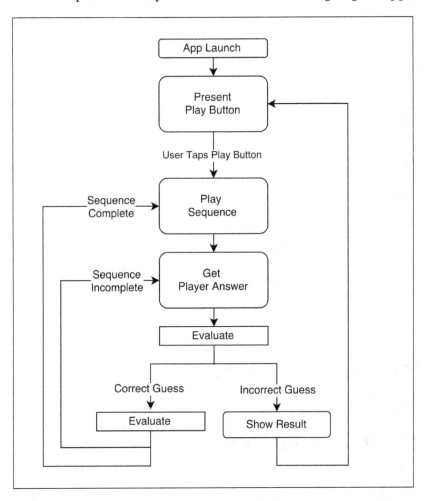

Now we're getting somewhere! Already we can get a pretty good idea of the the way in which the flow of the program suggests much of the code that we will be writing and this will be a valuable aid to ensuring that what we write is as clear and concise as we can get it, and that we only have to write it once.

We are now ready to think about how we will structure our app.

# App requirements

We will divide our app into three parts according to the program design principles espoused by Apple for writing iOS apps, the so called **Model-View-Controller** pattern.

- The **Model**, in this case, will be the actual game logic of generating the sequence, evaluating the player's guesses and deciding when the game is over.

- The **View** is simply the screen with which the player engages, which we will create in Interface Builder; it includes all of the UI elements we will need to interact with the player.

- The **Controller** will facilitate communication between the two by providing methods to present information to the player, gather player input, create timers, and set properties of the view to reflect the state of the app.

This separation of concerns is the bedrock of object orientated programming. It provides a pattern for designing robust apps and producing code that is both re-usable and easy to maintain. An expansive explanation of MVC is beyond the scope of this book, but the reader is encouraged to gain as much familiarity with the concept as possible.

# Setting up the project

For this app, our setup process will closely resemble that of the previous chapter.

# Create the Xcode project

To create the Xcode project, follow the given instructions:

1. Select **File | New | Project**, and from the templates select **iOS App With WatchKit App.**

2. Name the project **C-Quence** and deselect the Notification scene and any other scenes that may be preselected, we won't need them here.

The project window should now be looking similar to this:

# Create Required Classes

Looking at the project navigator on the left, we see that inside the **C-Quence** project is a folder also named **C-Quence**, which contains files that are relevant to the iPhone companion app that we will not be needing to change in this chapter. The next two folders are all about the **WatchKit Extension** and **WatchKit App**. We can see here that Xcode's default project structure adheres very closely to the MVC pattern described above.

The MVC's View is the `.storyboard` file that the template creates for us. Similarly, the Controller is already created, namely, the `InterfaceController.swift` file.

The only part of the pattern that we need to create from scratch is the Model, which is out next step:

1. Select **C-Quence WatchKit Extension** in the project Navigator and type *Command-N* to create a new file.

2.  Select **iOS | Source | Swift File**, as in the image below:

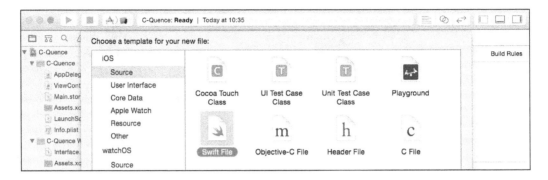

3.  Hit **Next** and name the new file `GameLogic.swift`, and be sure to check that the Group it is assigned to is the **C-Quence WatchKit Extension** and that the target selected is also **C-Quence WatchKit Extension** (since you selected the extension group in the Navigator before creating the file, these should be correctly preselected):

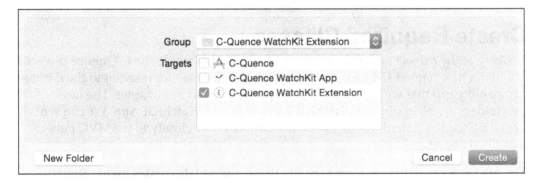

4.  Hit **Create**, as shown in the above image.

Capitalization is not a matter of taste.

There are certain conventions around when to use capitals and when not and these conventions can be critical in making your code easier to understand, either for other developers or for yourself later on.

Class names are always capitalized, as in the case of the `GameLogic` class we will create in a moment. Using further capitals for complete words in the name, such as the Logic in `GameLogic` is called camel case naming and is a great aid to clarity.

As we see above with `GameLogic.swift`, file names are also capitalized (but not the file type suffix!).

Variable names are written in uncapitalized camel case, for example `myVariableName`; this includes everything that is declared with the `var` and `let` keywords, as well as function argument names.

We now have all the files needed for a simple but well-structured app project that follows the Model-View-Controller program design pattern. It is a good idea to have some structure like this, however imperfect and incomplete it may be, before we start to dive into the coding and immerse ourselves in detail. Anything that we have missed, or that turns out quite differently to the way we expect, we can add and alter as we go along. Remember, a plan that may need revising is way better than no plan at all.

# Building the interface

The interface we will be building here makes use of the 'bordered' buttons we created in the previous chapter.

We need four such buttons, and we will use a hierarchy of *Group* objects to lay them out. Most of this layout method falls into the *drag-drop-set-value-repeat* category, but that does mean we'll be glad of the clear outlines we have already set out earlier.

In the project Navigator, select the `Interface.storyboard` file from the **C-Quence WatchKit App** group, and if the Attributes Inspector pane is not open, click *Command-Option-4*.

This is how the Group hierarchy that we are aiming to construct will be organized:

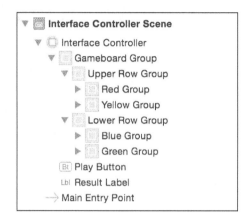

But it doesn't look that way yet:

We will drag a Group object into the Interface Controller, to be named **Gameboard Group**, that will contain all the buttons and groups visible during game play. Inside that we will organize two groups, the *Upper* and *Lower Row* groups, each of which will contain two color groups side by side. A button (not shown in the image above) will be placed in each color group.

# Set up the Group hierarchy

For setting up the Group hierarchy, follow the given steps:

1.  To save ourselves a lot of repetition, we will start by creating the **Gameboard Group**, then the **Upper Row Group**, and then the **Red Group** (and its button), and all the rest will be copying and pasting, and some renaming.

2.  Drag a Group object from the Objects Library (*Command-Option-Control-3*), select it in the Document Outline, and hit *Return* to give it the name **Gameboard Group**.

3.  In the Attributes Inspector:
    ○   Set the **Layout** property to **Vertical**.
    ○   Set **Spacing** to **Custom** and give it a value of **0**.
    ○   Set **Size Width** and Height to **Relative To Container**.

4. The Attributes Inspector should look like this:

5. Now drag a Group object into the **Gameboard Group** and name it **Upper Row Group**.

6. Set its **Insets** to **Custom**, and for each entry, **Top, Bottom, Left**, and **Right**, enter a value of **2**. You may like to tweak these values later on to your own tastes, but they are reasonable enough for the time being.

7. Set **Background Color** to **Black Color**.

8. Set the **Alignment**'s **Horizontal** and **Vertical** setting to **Center**.

9. Now set **Size Width** to **Relative To Container**. Do the same for **Height** – set it to **Relative To Container**, but give it a value of 0.5, meaning that it will always resize to be half the height of its containing view (in this case, the **Gameboard Group**), as pictured below:

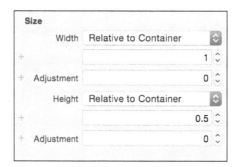

10. Next, drag a Group object into the **Upper Row Group**, and name it **Red Group**.

11. Set **Insets** to **Custom**, again with a value of **2** for each of the four insets.

12. Set the **Background Color** to a red that you find appealing.

13. The **Radius** property we will set to a **Custom** value of **8**, and this value you may well choose to alter once we see the UI on screen.

14. Set **Alignment Horizontal** and **Vertical** to **Center**.

15. Set the **Size Width** to **Relative To Container** with a value of **0.5**, set **Height** to **Relative To Container**.

## Add a button

Finally, drag a Button into the **Red Group**, erase its **Title**, since we have no need of a text title, and set its **Background Color** to **Black Color**.

- Set its **Size** Width to **Relative To Container**.

- Set its **Height** to **Relative To Container** but with an **Adjustment** of **-2** (that's a minus sign, not a hyphen!).

# Copy and paste

This is where the Document Outline really shines. Select the **Red Group.**

- Key in *Command-C* and then *Command-V* and you'll have the next color group automatically placed below the **Red Group** in the Document Outline, which in turn means it will be drawn second on the screen, and therefore to the right of it. Rename it **Yellow Group** and change its **Background Color** to something yellow, and we're done with the upper row.

- Select the **Upper Row Group** in the Document Outline, hit *Command-C* and then *Command-V*, and rename the new Group **Lower Row Group**. Inside that group we will need to rename the color groups to **Blue Group** and **Green Group** and change their **Background Color** properties to something appropriate. One might suggest blue and green.

Your Document Outline and interface should now look like these:

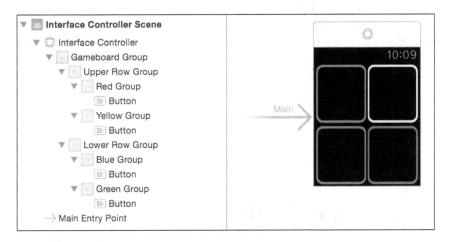

Select **Gameboard Group** in the Document Outline and set its **View | Hidden** setting (below the **Alpha** setting) to selected (ticked) in the Attributes Inspector.

It promptly disappears from the interface, so why did we do that? The game's interface is actually very simple, either we are looking at the gameboard that we have just built, or a play button (and possibly results label), and, rather than create different views, it is much simpler to hide whichever of the two 'views' that we don't need. At the launch of the app, we want to see the play button, but not the gameboard. So we hide it.

Setting the hidden property to `true` conveniently allows us to drag UI objects from the library onto (what is now) an empty view in Interface Builder, so let's do that now. Drag a Label object into the interface, and name it **Result Label**. Set its **Text Alignment** to **Center** and the **Lines** property to **0**. This will allow the label to show as many lines of text as is necessary, in case we decide to present something like "Hey, awesome, you just scored an amazing score of 128" when we show the user his result.

Set both **Horizontal** and **Vertical Alignment** to **Center**, and **Width** to **Relative To Container**. We'll set **View | Hidden** to `true`, since we don't need the result label when the app first launches. Now drag a Button into the (once again empty) interface, change its **Title** to **Play**, and set both **Horizontal** and **Vertical Alignment** to **Center**.

You're done! The interface is now complete. You could hit **Run**, but there won't be much to see apart from a **Play** button that doesn't do anything. First we need to create a game logic class and populate the `InterfaceController.swift` class with some code. Since the `InterfaceController.swift` code will need to call methods of the game logic class, we will implement the latter first. You'll see that the game's actual logic is really very simple, and this class will comprise little more than two dozen lines of code.

# GameLogic

Although we have created the file `GameLogic.swift` file, we have not actually created the class yet (this is different to Objective C)

## Create the GameLogic class

Below the `import Foundation` statement that is part of the template, add the following code:

```
import Foundation

class GameLogic {

}
```

# Plan the class

We will create a class that will encapsulate the code that deals with the game itself in isolation from the user interface. The GameLogic class doesn't need to know anything about interactions with the user, that is something that will be taken care of by the InterfaceController class, so let's first think about what we will need it to do, so that we can start to plan which methods we will need to implement.

We need it to do the following:

- Create and maintain a sequence of colors and add a random color to it when required
- Evaluate whether a player's tap on a color is a correct answer
- Provide information as to whether the game is still in play or finished
- Clear the data that collects during a game, in preparation for a new round

From these methods we can estimate that we will need to maintain at least one variable:

- A sequence property, which will be an ordered collection of colors

# Create the class's interface

Setting out our requirements like this has almost fully defined what the outside world needs the code to do. Not how, it's true, but that comes a little later.

We have effectively defined the class's external interface (not to be confused with the app's user interface that is something quite different), through which other parts of the app will communicate with it.

# Define some enums

In order to keep the code easy to read and safe to use, we will define some enums. We need to do this outside of the class itself, because the InterfaceController class will also need access to them. Add the following code directly after the import statement, but before the class definition:

```
import Foundation

enum Color {
    case Red, Yellow, Blue, Green
}
enum GuessResult {
```

```
        case GuessCorrect, GuessWrong, GuessComplete
}

class GameLogic {
}
```

Traditionally, enums have been a way to give names to integer values to make them more readable, but Swift has gone several steps further and dispensed with the idea that an `enum` needs some underlying numerical value. If you declare a variable's type to be an `enum` Type, like the above Color, the compiler will restrict you to those values, `.Red` and so on, and those values only. A method that is declared to return a Color will only return a Color and not some arbitrary integer (other convenient benefits include the fact that a switch statement needs no default once all the `enum` values are dealt with, as we'll see later).

So we now have four `Color` values, as we would expect, but why the third `GuessResult` value, `GuessComplete`?

When the `InterfaceController` asks the class to evaluate the player's answer, we can provide one of three possible scenarios; the guess is correct, but the sequence of guesses is not yet complete, the guess is wrong, or the guess is correct and the sequence has been correctly completed. Thus we save ourselves an extra call from the interface asking whether the guessed sequence is complete or not.

## Stub the methods

We can now `stub` the methods we'll need, which means creating them, and in some places, adding placeholder code to avoid compiler error messages. (Often we need to stub some actual functionality in the methods, but we will not need to do that here.)

Functions or methods?

Without going into an exhaustive definition of functions and methods worthy of a computer science degree course, it suffices to say here that when a function is coded as part of a class, it is referred to as a method of that class. If a function is defined outside of a class, it is called a function. In this app we have no functions outside of classes.

Add the following code inside (!) the `GameLogic` class:

# Extend the sequence

First, add a method that will extend the colors sequence by one randomly generated color:

```
func extendSequence() {
}
```

# Evaluate

Next we will need a method that evaluates whether the player has guessed correctly or not. We include a stubbed return value (that will be replaced later) in order to keep the compiler from warning us (with a very dramatic looking red circle and exclamation mark) that the method should return a `GuessResult` value.

```
func evaluateColor(color: Color) -> GuessResult {
        return .GuessCorrect
}
```

# Clear

Finally, when the game restarts, we'll need to clear the game of any variable values we have set in the course of play:

```
func clearGame() {
}
```

# Define properties

We will need a variable in which to store the sequence. The obvious type here would be an `Array` of `Color` values. We will initialize it to be empty at the beginning of the app's lifecycle. Add the following code above the `extendSequence` function:

```
class GameLogic {

    var sequence: [Color] = []

    func extendSequence(){
...
```

Our `GameLogic` class is now a reflection of the requirements we laid out earlier, and we can be confident that we have constructed a rugged frame to which we can add the logic that will make it perform its duties.

# Check your code

Your code should now look like this:

```
import Foundation

enum Color {
    case Red, Yellow, Blue, Green
}
enum GuessResult: Int {
    case GuessCorrect, GuessWrong, GuessComplete
}

class GameLogic {

    var sequence: [Color] = []

    func extendSequence(){
    }

    func evaluateColor(color: Color) -> GuessResult{
        return .GuessCorrect
    }

    func clearGame(){
    }
}
```

Check that there are no compiler errors or warnings.

All good? Splendid! On to the Interface Controller.

# Interface Controller

Next we go through the same process with the interface controller, namely planning out what we think it will need to do (remembering that an incomplete plan is good enough at this stage) and stub the methods that we expect to need.

## Planning the interface

The `InterfaceController` class will act as the CONTROLLER of communication between the user interface – the VIEW – and the actual implementation of the game – the MODEL. It needs to inform the player what the game is doing and inform the `GameLogic` class what the player is doing.

So, very roughly, the controller will need to do the following:

- Accept the player's instruction to start the game
- Flash the colors in sequence
- Accept player input
- Enable and disable user input as required
- Show results when a game is over
- Start a new game

## Define Outlets to the View

Add the following declarations to the top of the InterfaceController class (not the top of the InterfaceController.swift file):

```
class InterfaceController: WKInterfaceController {

    @IBOutlet var playButton: WKInterfaceButton!
    @IBOutlet var resultLabel: WKInterfaceLabel!

    @IBOutlet var gameboardGroup: WKInterfaceGroup!
    @IBOutlet var redGroup: WKInterfaceGroup!
    @IBOutlet var yellowGroup: WKInterfaceGroup!
    @IBOutlet var blueGroup: WKInterfaceGroup!
    @IBOutlet var greenGroup: WKInterfaceGroup!


    . . .
```

Without the @IBOutlet, Xcode will not enable us to easily connect UI objects to the code with a simple drag of the mouse.

These outlets (once we connect them) give us access to the UI elements whose properties we will need to change in some way during the game:

- We will need to hide/show the **Play Button**, the **Result Label**, and the **Gameboard Group** as required.
- The color groups being inside the **Gameboard Group**, will be hidden automatically, but we will also need to flash their colored 'borders' in sequence and so we need references to them too.

# Connect the UI with the Outlets

With the `Interface.storyboard` visible in the editor window, make sure the assistant editor is showing, as can be seen in the image below (if it's not, hit *Command-Option-Return*). Use the Document Outline to Control-drag from the Play Button to its corresponding `@IBOutlet var playButton` outlet in the code:

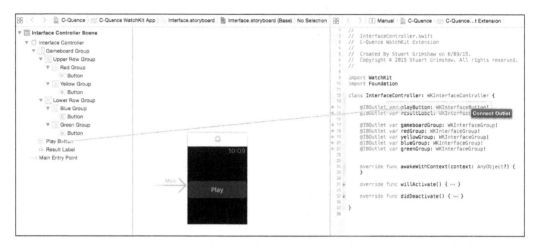

In the same way, make the following connections:

- **Result Label** to the `@IBOutlet var resultLabel` outlet.
- **Gameboard Group** to the `@IBOutlet var gameboardGroup` outlet.
- **Red Group** to the `@IBOutlet var redGroup` outlet.
- **Yellow Group** to the `@IBOutlet var yellowGroup` outlet.
- **Blue Group** to the `@IBOutlet var blueGroup` outlet.
- **Green Group** to the `@IBOutlet var greenGroup` outlet.

# Stub some preliminary methods

Now we'll sketch out which methods we expect we'll need the controller to perform. All of these methods belong inside the curly braces of `class InterfaceController: WKInterfaceController`.

Already provided by the template is the `awakeWithContext` method. This we can leave unaltered until we come to fill out the method's code:

```
override func awakeWithContext(context: AnyObject?) {
}
```

## Start the game

The app launches with the **Play Button** visible, so we'll add a method to respond to the player tapping it, which we will connect later (hence the @IBAction):

```
@IBAction func playButtonTapped() {
    }
```

When we come to write the logic, we will resist the temptation to stuff a load of code into the playButtonTapped method and instead directly call a startNewGame method:

```
    func startNewGame() {
}
```

> Many developers, including this author, like to keep the methods called by UI objects (such as @IBAction func playButtonTapped) as slim as possible and restrict them to containing a single method call.
>
> In the case of playButtonTapped this is a single call to StartNewGame. The reason for this is that we may need the combination of calls initiated by a button tap at other points in the app that are not associated with that particular user action. The code we put into StartNewGame will be accessible through a method call that reflects *what* we want to happen, rather than *how* it was triggered.
>
> We could, in fact, make a call to playButtonTapped at any point in the code, but it's a bad habit, leading to confusing code. This method is reserved for the UI button only.

Referring to our user story above, we will now need to play the sequence of colors (at this point, it will be a sequence of one):

```
    func playSequence(){
}
```

This in turn will require a flashColor method:

```
    func flashColor(color: Color, duration: Double) {
}
```

We won't call this method directly; it will be set as the method to be called by an NSTimer object that will be started by playSequence. The timer may need to do more than just flash a color, so instead it will call a timerFired method, that will in turn call flashColor and any other methods necessary.

```
    func timerFired() {
}
```

## Accept guess

We create an IBAction for each button (also to be connected shortly):

```
    @IBAction func redButtonTapped {
}
    @IBAction func yellowButtonTapped() {
}
    @IBAction func blueButtonTapped() {
}
    @IBAction func greenButtonTapped() {
}
```

Each of these will call the same method, `colorButtonTapped`, which takes as its argument the relevant `Color`:

```
    func colorButtonTapped(color: Color) {
}
```

## Game over!

The last method we will add to `InterfaceController` is the one that is called when the player has guessed incorrectly and the current round of the game is thus at an end. This method we will later populate with code to display the score and the **Play Button** as well as prepare the app for the next round. In order to know what the score is, it will take an `Int` as its single argument.

```
    func endGame(result: Int){
}
```

## Check your code

Your `InterfaceController.swift` file should now look very similar to this:

```
import WatchKit
import Foundation

class InterfaceController: WKInterfaceController {

    @IBOutlet var playButton: WKInterfaceButton!
    @IBOutlet var resultLabel: WKInterfaceLabel!

    @IBOutlet var gameboardGroup: WKInterfaceGroup!
    @IBOutlet var redGroup: WKInterfaceGroup!
    @IBOutlet var yellowGroup: WKInterfaceGroup!
    @IBOutlet var blueGroup: WKInterfaceGroup!
```

```swift
@IBOutlet var greenGroup: WKInterfaceGroup!

override func awakeWithContext(context: AnyObject?) {
}

@IBAction func playButtonTapped() {
}

func startNewGame() {
}

func playSequence(){
}

func flashColor(color: Color, duration: Double) {
}

func timerFired() {
}

@IBAction func redButtonTapped() {
}
@IBAction func yellowButtonTapped() {
}
@IBAction func blueButtonTapped() {
}
@IBAction func greenButtonTapped() {
}

func colorButtonTapped(color: Color) {
}

func endGame(result: Int) {
}

override func willActivate() {
    super.willActivate()
}

override func didDeactivate() {
    super.didDeactivate()
}

}
```

Once again, check that your code is provoking no adverse reactions from the compiler and, with that done, we are ready to connect the UI objects to the code.

# Hook up the UI with the Outlets

To hook up the UI with the Outlets, follow the given instructions:

1.  Once again we will *Control-drag* from the document outline into the code of `InterfaceController.swift` in the assistant editor.

2.  Drag from the **Play Button** to the `@IBAction func playButtonTapped()` method, as shown below. In this way the button is connected to the IBAction that will be called when it is tapped.

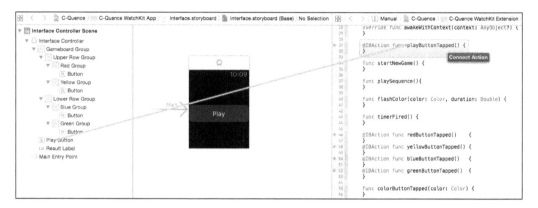

3.  Do the same thing for the button in each of the four color groups. Be sure to click on the button, not the group, or Xcode will not allow the connection.

You may have wondered why we have not given the buttons names, or attached them to outlets. This is because we will not need to refer to the buttons in the code—as long as they trigger the required `IBAction`, that's all we need of them. They will be hidden along with their containing groups and their appearance doesn't change.

If you ever need to check on a UI object's connections to outlets and actions, you can do so by Control-clicking it, after which you will be presented with a summary of all its connections. It is also here that it is easiest to delete a connection or action when the need arises, as shown below:

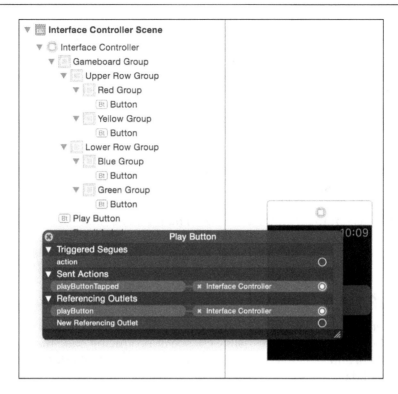

(The astute reader may also notice the Triggered Segue fields. These we will cover in a later project.)

# Run the app

Now, that was a long stretch of coding and, if you were to run the app now, you might feel a little short -changed. You'll see a solitary **Play** button, which does absolutely nothing at all. Or at least, appears to do nothing. Of course, what it actually does is trigger the `playButtonTapped` method, which at the moment is profoundly empty of code.

Before we go any deeper into the code, let us look back at what we have done so far, how we went about it, and why we did it. One of the most common mistakes made by less experienced developers (and no small number of experienced ones) is to dive into the detail too soon. Before you know it, you have spent an inordinate amount of time going into the minutiae of some small area of the app, one which may perhaps never even be used once the broader context becomes clear. Most or all of what you have written may need to be rewritten and much of what you write afterwards may be adversely influenced by the psychological weight of these first lines of code.

In going through the steps you have taken, you have approached the code first from a global perspective, from which you can see what you want your app's code to achieve, without having to worry about how it will do it. Before you can add the windows and staircases to a house, you need to have the form of the house clear, with a strong framework in place (analogies to building and architecture abound in the literature of software development, and the tome you have before you will be no exception).

Your second step was to go one level deeper, and think about what your code will have to do, in terms of the user's experience and the flow of the program. Vague structures start to crystallize into pseudocode that in turn starts to produce an ad hoc list of methods, some of which suggest other methods that may be needed. If a method turns out to be unnecessary, then no harm done—deleting is cheap.

It is worth repeating that an imperfect plan is all we needed at this stage.

# Summary

In this chapter you have laid the groundwork for the rest of the app and done so in a way that gives you the best possible chance of producing code that is robust, easy to maintain, and easy to comprehend when returning to it after six months on a surfing holiday.

Specifically, you have learned to plan the app according to a projected user story, applying the Model-View-Controller design pattern, and then set up the project according to that pattern; you have used Group objects to layout the user interface in Interface Builder and you have prepared the Interface Controller class's methods. You have also implemented some best programming practices by encapsulating the game logic into a class of its own and using enums to make your code easier to write and clearer to read.

In the following chapter we will flesh out the methods we have already declared, and begin to run, evaluate, and improve the code; we will add the ability to navigate to other screens, and enable the user to customize the app experience and store preferences data. We will also add images to the watch interface and enable the Watch and the iPhone to communicate and share data.

# 4
# Expanding on C-Quence

Now that we have a strong framework in place, we can begin to turn those empty functions into fully implemented methods, after which *C-Quence* will be a fully functioning watchOS app.

You will probably be delighted to discover just how easy this is, having reached a point at which we have already decided both how the app is structured at a high level and roughly which methods we need our two classes to provide. The basic app may *look* only half finished, but it is, in fact, much further than that.

## Implementing the methods

Take a deep breath, we don't run the code for a while!

We will start with our `InterfaceController` class. This is where the user's engagement with the app begins, and this is where we will begin to endow the classes with life.

When the app launches, it automatically loads the view that is designated as the **Main Entry Point**, identified in Interface Builder by the **Main** arrow, as shown in the following figure:

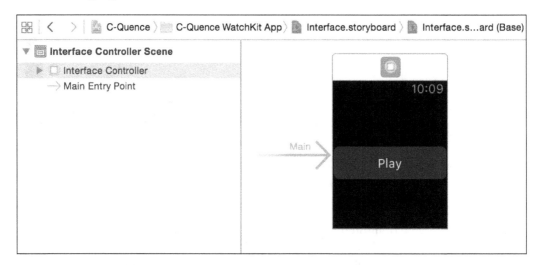

In our case, this main entry point is the InterfaceController class.

# The InterfaceController class

In addition to performing duties of its own, InterfaceController creates two objects on which it must call. Firstly, an instance of our GameLogic class and secondly, an instance of NSTimer with which we will control the timing of the flashing colors.

In contrast to some languages, in Swift we can both declare these properties and instantiate them at the same time. Add the following code to the InterfaceController class below the IBOutlet declaration:

```
@IBOutlet var greenGroup: WKInterfaceGroup!

var gameLogic: GameLogic = GameLogic()
var timer: NSTimer = NSTimer()

override func awakeWithContext(context: AnyObject?) {
```

Often, we do not need to override a class's init method, if, as in this case, all the class's properties are initialized at declaration. We are effectively saying in one line of code *'This class has a property X and its initial value must be set to Y'*. This is a pretty neat and simple way to write initialization code!

# awakeWithContext

When the class has finished loading, our next chance to interact with the app comes when it calls the `awakeWithContext` method from the `WKInterfaceController` class (comparable with UIKit's `viewDidLoad` method from the `UIViewController` class), which it does without any help from us. This is our chance to set up any properties of the `view` to a suitable initial state.

The initial state we will set here is the value of the button border colors `opacity` property or `alpha` value. But before we do that we will create some constant `alpha` values.

We achieve the flashing effect on screen by changing the brightness of the color groups background color; by changing its transparency, which is to say its `alpha` value, it will appear to increase its intensity. This way of flashing the colors means we will be referring in the code to two levels of transparency, or two `alpha` values, and it is a good idea to define these as `constant` values (using the `let` keyword) that we can then use in our code (instead of repeatedly writing `0.5`, and having to change all those occurrences later, if we decide on a different value).

 An alpha value of 1.0 means completely opaque, 0.0 translates to completely transparent, i.e. invisible.

Add the following code directly beneath the `import` statements:

```
import WatchKit
import Foundation


let fadedColorAlpha: CGFloat    = 0.4
let flashColorAlpha: CGFloat    = 1.0
```

If, later, you should choose some other value for the `fadedColorAlpha` value, this is the only place you'll need to do it.

Now we can use these constants in the `awakeWithContext` method, adding the following code:

```
override func awakeWithContext(context: AnyObject?) {
    super.awakeWithContext(context)
    for group in [redGroup, yellowGroup, blueGroup, greenGroup] {
        group.setAlpha(fadedColorAlpha)
    }
}
```

All we are doing here is creating an impromptu `Array` of the four color groups and then enumerating through that array, setting each group's `alpha` property to the `fadedColorAlpha` we defined previously. Thus, each button 'border' looks dulled down against the black background of the Upper and Lower Row Groups.

Our view is finished and ready for play. An excellent reason, you'll no doubt agree, to implement the code that gets called when the player taps the **Play** button.

## playButtonTapped

As we saw earlier, it is a good idea to keep `IBAction` methods really slim. This one simply needs to call another method — `startNewGame` — that will, well, start the game.

 If, in some future version of the game, you should enable the player to start the game by performing a backward somersault, you will be able to call the same code with your `@IBAction func userDidFlip()` method, or any other method deemed appropriate.

Add the following call to `playButtonTapped`.

```
@IBAction func playButtonTapped() {
        startNewGame()
    }
```

The player having indicated that he is ready to play the game, we can hide the **Play** button, unhide the `gameboard` itself, and make a call to the `playSequence` method:

```
func startNewGame() {
    gameboardGroup.setHidden(false)
    playButton.setHidden(true)
    playSequence()
}
```

## Declare additional constants

Since the `InterfaceController` duties include the flashing of the buttons, we need to make some decisions around the speed and the duration of the flashes. We will actually define two flash durations, but the use of the shorter one will become apparent later. One second seems to be a suitable time between the start of each flash.

Add these constants to the alpha values we defined earlier beneath the import statements:

```
let fadedColorAlpha: CGFloat    = 0.4
let flashColorAlpha: CGFloat    = 1.0
let longFlashDuration: Double   = 0.8
let shortFlashDuration: Double  = 0.3
let timerInterval: Double       = 1.0
```

While the sequence is flashing, we want to prevent any actions being triggered by the user tapping the buttons, so we'll add a Bool variable, or flag, to keep track of whether the buttons should call some method.

We also need to keep track of where we are in the sequence while it is being flashed, so we define an Int variable to store this information.

These var are added to the variable declarations below the IBOutlet declarations:

```
var gameLogic: GameLogic = GameLogic()
var userInputEnabled = false
var indexOfNextColorToFlash = 0
var timer = NSTimer()
```

# playSequence

Now we can flesh out the playSequence method by adding the following code:

```
func playSequence(){
    userInputEnabled = false//1
    indexOfNextColorToFlash = 0//2
    gameLogic.extendSequence() //3
    timer = NSTimer.scheduledTimerWithTimeInterval( //4
        timerInterval,
        target: self,
        selector: "timerFired",
        userInfo: nil, //5
        repeats: true)    }
```

Here is a detailed look at what we have added (see the comments, //1…etc. in the code above):

1. Before the sequence starts playing, we set userInputEnabled to false.

2. We want the sequence to play from the beginning, so indexOfNextColorToFlash is set to 0.

3. We make the call to extendSequence method from the gamelogic class to add a random color to the sequence.

4. We start timer with the scheduledTimerWithTimeInterval call. We set the timerInterval argument to be the timerInterval constant we defined earlier, set the InterfaceController instance to be the target, using the self keyword, so that the timer knows from which instance it should call the selector, the selector here being timerFired.

5. We do not need to make use of the userInfo dictionary, but we do need to set repeats to true.

## timerFired

Finally, we get to flash something on the screen! Add the following code to our timerFired method:

```
func timerFired() {
    let colorToFlash =
    gameLogic.sequence[indexOfNextColorToFlash] //1
    flashColor(colorToFlash, duration: longFlashDuration) //2
    if indexOfNextColorToFlash < gameLogic.sequence.count - 1//3
    {
        indexOfNextColorToFlash++
    } else {
        timer.invalidate() //4
        userInputEnabled = true
    }
}
```

The comments in the code are as follows:

1. We read the color from the current index of the sequence array.

2. We pass this Color to the flashColor method for it to begin its animation that produces the flashing effect along with our previously defined longFlashDuration constant, which you may wish to experiment with later.

3. If we have not reached the end of the sequence, we increment the index of the sequence array to be accessed the next time the timerFired method is triggered (which will happen automatically, since we have set its repeats property to true).

4. Otherwise we kill the timer by calling its invalidate method, and set the userInputEnabled flag to true, in anticipation of the player's input.

# flashColor

Here is the code for the `flashColor` method:

```
func flashColor(color: Color, duration: Double) {
    let group: WKInterfaceGroup
    switch color { //1
    case .Red:
        group = redGroup
    case .Yellow:
        group = yellowGroup
    case .Blue:
        group = blueGroup
    case .Green:
        group = greenGroup
    }
    group.setAlpha(flashColorAlpha) //2
    animateWithDuration(duration) { //3
        group.setAlpha(fadedColorAlpha)
    }
}
```

The comments in the code are as follows:

1. We select the color group appropriate to whichever `color` argument was passed to the method.
2. We set the `alpha` value of that group to our constant value `flashColorAlpha`, making it seem brighter and thus appear to light up.
3. We use `animate` method from the `WKInterfaceController` class to have the `alpha` value fade back to its original value, with a duration set to the `duration` argument that was passed to the method.

# redButtonTapped & Co

Once the color sequence is done flashing, the app just waits. It will do nothing until the user taps one of the buttons.

True to form, each `IBAction` here only contains a single method call (to `colorButtonTapped`) passing an appropriate `color` value as its only argument.

Add the following code to the four IBAction methods to which we have already connected the buttons of the user interface:

```
@IBAction func redButtonTapped() {
    colorButtonTapped(.Red)
}
@IBAction func yellowButtonTapped() {
    colorButtonTapped(.Yellow)
}
@IBAction func blueButtonTapped() {
    colorButtonTapped(.Blue)
}
@IBAction func greenButtonTapped() {
    colorButtonTapped(.Green)
}
```

# colorButtonTapped

Add the following code to the colorButtonTapped method:

```
func colorButtonTapped(color: Color) {
    if userInputEnabled { //1
        flashColor(color, duration: shortFlashDuration) //2
        let guessResult = gameLogic.evaluateColor(color) //3
        switch guessResult {
        case .GuessCorrect: //4
            break
        case .GuessWrong: //5
            endGame(gameLogic.sequence.count - 1)
        case .GuessComplete: //6
            playSequence()
        }
    }
}
```

The comments in the code are as follows:

1.  We check to see if the userInputEnabled flag is set to true. If it is not, then we are not currently accepting input from the player, and the method does nothing.

2. If the flag is `true`, we call the `flashColor` method, passing along the `color` argument that was passed into this method call. We also pass the constant `shortFlashDuration` as the second argument; this shorter duration is to acknowledge the player's button tap. The game seems somewhat snappier with the shorter duration.

3. We pass the `color` tapped by the player as the argument to `evaluateColor` method from `gameLogic` class, which returns a `GuessResult` value.

4. If the guess is correct, but the sequence is not yet complete, we do nothing. We wait for the player's next guessed color.

5. If the guess is incorrect, then the game is over. The length of the sequence at the player's last successful completion was one less than it is on this unsuccessful completion. This `Int` is passed as the `result` argument of the `endgame` method.

# endGame

Add the following code to the `endgame` method:

```
func endGame(result: Int) {
        gameboardGroup.setHidden(true) //1
        resultLabel.setHidden(false) //2
        resultLabel.setText("Not bad! You scored \(result)")
        playButton.setHidden(false) //3
        playButton.setTitle("Play Again")
        gameLogic.clearGame()
}
```

The comments in the code are as follows:

1. Now that this round is over, we want to hide the `gameboard`.

2. We show the `resultLabel`, with its text reflecting the `result` argument. At the moment, the *Not bad...* message is a little static, but this we will improve later.

3. Show the **Play** button to the player. While we're at it, we may as well update the `Title` property of the button to reflect the fact that this is not the first round.

4. Tell `gameLogic` to clear the game, which we shall implement next.

# GameLogic

Now it's time to implement the GameLogic class.

## Init the sequence array

When the InterfaceController method creates an instance of the GameLogic class, we first need to create a container for the sequence of colors that will be generated as the game progresses. We declare an Array of type [Color] and initiate it to be empty with the following code, which belongs at the beginning of the GameLogic class definition:

```
class GameLogic {

    var sequence: [Color] = []

    func extendSequence() {
```

## Extend the sequence

The first of the methods we will implement in this class will add a randomly generated Color to the sequence array.

```
func extendSequence() {
    let randomInt = Int(arc4random_uniform(4)) //1
    let nextColor: Color =
    [.Red, .Yellow, .Blue, .Green][randomInt] //2
    sequence += [nextColor]
}
```

The comments in the code are as follows:

1. We generate a random integer between 0 and 3 using the standard library's elegantly named arc4random_uniform function.

2. We use this integer to select a color from a Dictionary of the four valid values of Color and add it to the sequence array.

## Evaluating the user's input

When we evaluate whether the user's last guess is correct, we need to have kept track of which position in the sequence it is that's being guessed. This Int property will be declared when the instance of GameLogic is initialized, with its initial value set to 0.

Add the following code to the class's var declarations:

```
var sequence: [Color] = []
var nextAnswerIndex: Int = 0
```

Now that we have done that, we are ready to fill in the evaluateColor method with the following code:

```
func evaluateColor(color: Color) -> GuessResult {
    if color != sequence[nextAnswerIndex] { //1
        return .GuessWrong
    } else {
        if nextAnswerIndex < sequence.count - 1 { //2
            nextAnswerIndex++
            return .GuessCorrect
        } else {
            nextAnswerIndex = 0//3
            return .GuessComplete
        }
    }
}
```

The comments in the code are as follows:

1. If the guess does not match the color at the appropriate position of the sequence array, we return .GuessWrong.

2. If the guess is correct, we test whether it was the last in the sequence. If it is not, we increment the index of the array with which the next guess will be compared, and return a value of .GuessCorrect.

3. If the correct guess was the last in the sequence as it currently stands, we will return a value of .GuessComplete, but before that we need to reset the nextAnswerIndex variable to 0 in preparation for the next round.

## clearGame

When an instance of the GameLogic class receives the instruction to clear the game, in preparation for starting a whole new game, it simply needs to reset the sequence and nextAnswerIndex variables to the values with which they were initiated:

```
func clearGame() {
    sequence = []
    nextAnswerIndex = 0
}
```

# Test it

That's it! Take a deep breath, you are about to run the code and see how it works. Check that you have no compiler warnings or errors by typing *command-4* to show the **Issue Navigator**. The list of issues should look healthily empty:

# Build and run the app

Go ahead, hit **Run**.

You will be presented with the **Play** button, and on tapping it you will be able to play complete rounds of the game.

# Testing and tweaking

One of the hardest things in developing software is testing its functionality to the limit. There is something inherently difficult about trying to provoke your own code into failing your own tests, but it is absolutely imperative that you cover as many as possible of the scenarios that the app could come up against. Users are so unpredictable.

# The first test

With a simple app such as this one, this is not such an insuperable task. Play the game over and over, trying to catch not only the usual sequence of actions that you expect your users to go through, but also all those niggly edge cases, such as:

- User fails at the first guess
- User taps incessantly on the screen
- User manages an incredibly long sequence (pen and paper might help – or just a lot of practice)
- User switches to a different app and then returns
- Anything else you can think of
- Anything that anybody else can think of

# The first bug

Now, if you have tested your creation on a real Apple Watch, you may have noticed that we have a problem with our game. The screen will only stay active for about 15 seconds if the user makes no interactions with it, then it goes black. Your app sleeps. And thus is the battery life of the Apple Watch significantly improved, so this is a Good Thing. Really.

Not got a watch yet? That's not going to be a huge drama in this chapter, but in order to actually build, test, and ship apps to the App Store, you will most certainly have to invest in the hardware. Our newly discovered bug is just one of innumerable differences between the behavior of Watch Simulator and the real thing.

But what to do? Well, let's take on board the fact that the Apple Watch is about brief interactions. Not all as brief as checking the time or reading a text, certainly, but we're not talking about an iPhone game here that a user will happily play for hours while waiting for a delayed flight at the airport. We can use this small hiccup as a spur to our ingenuity, and try to find a fix that sits well with the device that we are coding for – the Apple Watch.

# The first fix

So instead of looking for a way to prolong the illumination of the screen, we must find a way for our game to become more difficult without becoming longer.

What we will change is the speed at which the sequence is presented to the user. Beyond a certain number of steps, we will start to accelerate the pace, ensuring that the game's duration never exceeds a certain limit.

Add a property, `intervalFactor` to `InterfaceController` properties:

```
var gameLogic: GameLogic = GameLogic()
var userInputEnabled = false
var indexOfNextColorToFlash = 0
var timer = NSTimer()
var intervalFactor = 1.0
```

We must multiply the timing durations by this value throughout the game and decrease the value as the sequence gets longer, thus speeding up the rate of flashing and limiting the game's duration.

We will also add a constant that will define the maximum duration of a game. Let's play it safe and set this to 12 seconds (which is a few seconds short of the maximum time an app will stay active with no user interaction):

```
let fadedColorAlpha: CGFloat      = 0.4
let flashColorAlpha: CGFloat      = 1.0
let longFlashDuration             = 0.8
let shortFlashDuration            = 0.3
let timerInterval                 = 1.0
let maximumSequenceDuration       = 12.0
```

Now all we need to do is check how long a flash is going to last each time the sequence gets extended. Change the code of the `playSequence` method to the following:

```
func playSequence(){
    userInputEnabled = false
    indexOfNextColorToFlash = 0
    gameLogic.extendSequence()
    let maximumNumberOfFlashes
          = Int(maximumSequenceDuration / timerInterval) //1
    if gameLogic.sequence.count > maximumNumberOfFlashes{
        intervalFactor = //2
            maximumSequenceDuration / Double(gameLogic.sequence.
count)
    }
        timer = NSTimer.scheduledTimerWithTimeInterval(
        timerInterval * intervalFactor, //3
        target: self,
        selector: "timerFired",
        userInfo: nil,
        repeats: true)
}
```

The comments in the code are as follows:

1. We calculate `maximumSequenceDuration / timerInterval`, which gives us the maximum number of color flashes that we can achieve in the allowed sequence duration, and convert it to an `Int`, so that we may compare it to the length of the sequence, `gameLogic.sequence.count`.

2. If the length of the sequence exceeds that maximum number of flashes, we adjust the `intervalFactor` by setting it to the value of `maximumSequenceDuration` divided by the new sequence length (which has been converted to a `Double` to allow the calculation).

3. Now we can adjust the timer duration by our new `intervalFactor` property.

## Test again

Run the app again, and you will see that after a while, the pace of the game picks up and the colors are flashed at increasing speed.

 You may wish to set `maximumSequenceDuration` to something like `3.0` to make testing quicker — and easier on the brain!

But now there is another problem. There soon comes a point at which the individual flashes are not finishing fast enough to make them clearly distinguishable; for example, when there are three consecutive reds in the sequence. Clearly we must also modify the duration of the animations.

## Fix again

1. Change the following code in the `flashColor` method:

```
func flashColor(color: Color, duration: Double) {
    let group: WKInterfaceGroup

    . . .

    group.setAlpha(flashColorAlpha)
    animateWithDuration(duration * intervalFactor) {
        group.setAlpha(fadedColorAlpha)
    }
}
```

Note how we only need to change one line of code for both short and long animation durations, since they both call the same method, and the `duration` argument passed to that method is simply multiplied by the ever-decreasing `intervalFactor`.

Number types—You cannot do calculations in Swift with mixed numerical types, so you must use the number types conversion methods—Int (Double) for example, takes a Double as its argument and returns the highest Int less than or equal to that Double's rounded-down value.

## Test

Run the code again and check that the results are to your taste; you may want to try tweaking the duration constants to improve the subjective pace of the game.

But now we have a new problem; the second round of the game (after the player has made an incorrect guess) starts at this new increased speed.

## Fix

So when we start a new game, we need to reset the `intervalFactor` to its original value:

```
func startNewGame() {
    gameboardGroup.setHidden(false)
    playButton.setHidden(true)
    resultLabel.setHidden(true)
    intervalFactor = 1.0
    playSequence()
}
```

## Test again

Run the code again…well, you get the idea. If there is anything left that you feel you can improve on, go ahead; this is your work, and when you demonstrate it to someone, you want the little details to be right, right? Are the colors well balanced? How is the spacing of the color groups, the pace of the app after launch? The **Play** button to your liking?

Assuming you have ironed out the creases and got things just the way you want them (or close), we finally have the app in a state in which it fulfills both our original requirements as set out at the beginning of this chapter, and some further requirements we identified along the way as development progressed. Simple as it is, you now have a rugged codebase, structured for efficient development and future maintainability.

But why stop there? Well of course, you won't. There is a limitless number of steps we can take to increase the quality of the user experience. And many of the ideas you may have will depend on the watch communicating with its larger sibling, which we have not mentioned for some time: the iPhone.

# Communicating with the phone

The app that we have coded is in some senses completely autonomous — the iPhone companion app does nothing except install the watchOS app, after which the player never needs to engage with the phone. Which is kind of a shame given that the phone offers us some benefits that most apps are likely to be able to make use of. These include:

- Much greater storage capacity for resources
- Much greater processing power (and the battery life to do it)
- A much larger screen

While we will have little use in *C-Quence* for the first two of these, we will make use of the ease with which our users can give us information using the larger screen, information that we will only collect once, or at least not often.

As mentioned in *Chapter 1, Exploring The New Platform*, the Apple Watch is at one and the same time an independent device, and an extension of the iPhone, but we can also turn this relationship around and use the iPhone to augment the capabilities of the Watch.

To get input from the user, we will leverage the larger screen of the iPhone. And to do that, we'll need to get the phone and the watch talking to each other.

# Make it personal

We will add a little of the personal touch to our app by having it address the user by his name, one that he will enter into the iPhone (we won't expect anyone to enter text directly into the Watch), which will be transferred automatically to the watch.

## Getting the message across

There are several methods available to us to facilitate the transfer of data in both directions between the iPhone and its paired Apple Watch using methods supplied by Apple's `WatchConnectivity` framework. In this chapter, we will use some of its **Application Context** methods, which are possibly the simplest to implement, as well as being best suited to our requirements. We will sketch out these requirements first.

## Requirements

An extension to our app necessitates an extension to our user story.

The player will be given the opportunity of entering her name into the app from the view presented on the iPhone at launch.

 Since our focus here is very much on the watch, we will create only the bare minimum UI necessary for the user to enter a name. The reader is, of course, encouraged to add some individual flair to the iPhone UI.

The iOS app will then send the entered name to the watchOS app, which will store it in a form that makes it accessible after the app has terminated and re-launched.

## What classes will we need?

Since communication between the watch app and phone app is neither a part of the game logic, nor a part of interfacing with the user, we will create new, dedicated classes on both the iPhone and the Watch, which will take care of all aspects of inter-app data transfer using `WatchConnectivity`.

In addition to methods from `WatchConnectivity`, we will also add a custom method, `sendNameToWatch`, to the iPhone app.

# Preparing both apps to communicate

Using `ApplicationContext` methods from the `WatchConnectivity` class involves populating and sending a `Dictionary` object, the same type of object that you will probably have used in iOS apps already. These `Dictionary` objects must be in the form [`String :AnyObject`], which is specified by the `ApplicationContext` methods we will be using. Since both the watch app and the phone app will need to be using exactly the same keys (so in this case, the `String`) to access the values of the `Dictionary`, it is advisable to create a file of predefined constant String objects, which will save us the trouble caused by typing mistakes, meaning that one app sends, for example, [`"Color" : RedObject`], while the other looks for [`"color" : RedObject`] (note the capitalization), an easy mistake to make, and one that will cause your code to fail.

 Using String constants will, incidentally, get Xcode's code-completion feature working for us, as you'll see when you start to use them in your code.

To do this, we must create a file that is accessible to both apps, which means ensuring it is available to both the *C-Quence* and the *C-QuenceWatchKit* Extension targets:

# Create a Constants.swift file

Before we define the constants, we must create a Swift file for this purpose, with the following steps:

1. Type *command-N* to create a new file, using one of the Xcode templates

2. Choose **iOS | Source | Swift File**.

3. Hit **Next**.

4. In the following dialog box, name the file `Constants.swift`

5. Select the Group **C-Quence**, and tick both the **C-Quence** and **C-QuenceWatchKit Extension** targets, as shown below:

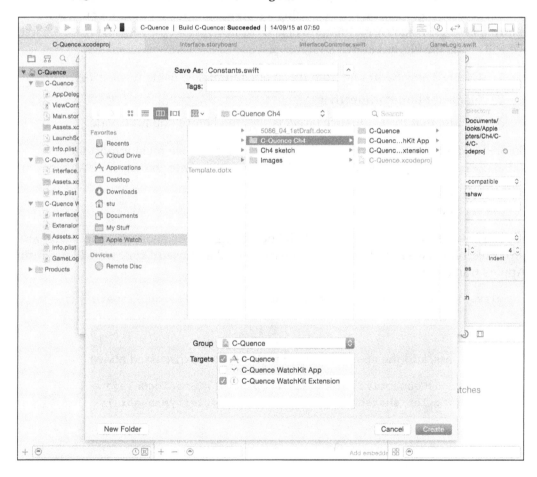

6. Select the location for the file, at the top level of your project, since this file will be used by more than one target, and hit **Create**.

## Define the constants

Add the following code to the new `Constants.swift` file:

```
let kPlayerName = "playerName"
```

# Create the iPhone Connectivity Manager

We must now turn our attention to the iPhone app.

The first thing we will do is create the class that will take care of all matters pertaining to data transfer between the two apps, which we will call **PhoneConnectivityManager**. There will only ever be one instance of this class (what would we do with two?), so we will make it a `singleton` object, by adding a `sharedManager` method that will always return this single instance, no matter when, or from where in the code, it is called. First, we will create the file, as below:

1. Create a new Swift File from the Xcode templates, using *command-N*.
2. Name it **PhoneConnectivity**.
3. Be sure that the **C-Quence** target is selected before you hit **Create**.

Before we define the `PhoneConnectivityManager` class, we must import the `WatchConnectivity` framework. Add the following code to the `import` statement:

```
import Foundation
import WatchConnectivity
```

We can now define the class, adding to it both the `NSObject` and `WCSessionDelegate` protocols, which we need to access all the really cool and useful stuff we will use from Apple's frameworks:

```
class PhoneConnectivityManager: NSObject, WCSessionDelegate {

}
```

Inside this class, add the `sharedManager` method, as we discussed above:

```
class PhoneConnectivityManager: NSObject, WCSessionDelegate {
    static let sharedManager = PhoneConnectivityManager()

}
```

Directly after the `sharedManager` method, add the following `init` code:

```
private override init() { //1
    super.init() //2
    if WCSession.isSupported(){ //3
        let session = WCSession.defaultSession() //4
        session.delegate = self //5
        session.activateSession() //6
    }
}
```

Let's take a look at what we're doing here:

1.  We mark the `init` method `private`, so that the only way to create an instance of the class is through the `sharedManager` method, thus ensuring that the single shared instance is returned every time.

2.  The `NSObject` protocol requires us to call the superclass's `init` method, the details of which need not concern us here (they're interesting but way out of scope).

3.  We check if the device that the app is running on supports `WatchConnectivity`; an iPad will return `false` when `isSupported` is called.

4.  If we are on an iPhone, we initiate an instance of `defaultSession`.

5.  This session will send messages to its `delegate` (this is why we added the `WCSessionDelegate` protocol), so we assign `self` to be that delegate.

6.  We have completed preparation of the session and we can now activate it.

With the `WatchConnectivity` session set up, we now need to create a method, to be called from other parts of the app that will actually send the player's name to the watch.

After the `init` method, add the following code:

```
func sendNameToWatch(playerName: String) {
    if !WCSession.defaultSession().paired { //1
        print("No paired watch")
        return
    }
    if !WCSession.defaultSession().watchAppInstalled { //2
        print("Watch app not installed")
        return
    }
    do { //3
        let context = [kPlayerName: playerName] //4
```

```
            print("Sent \(playerName) in ApplicationContext")
            try WCSession.defaultSession().updateApplicationContext
(context) //5
        } catch {
            print("applicationContext update failed")
        }
    }
```

Once again, let's break this down to understand exactly what we are doing:

1. We check if the phone is paired with a watch. If not, our job is done and we `return` early, logging this case to the console for possible debugging.

2. Similarly, we check if our app has been installed on the watch (the user can choose not to) and `return` if not, again logging to the console.

3. We will wrap the call to send the message in a `do catch` block and prepend it with the `try` keyword, which is Swift's error handling pattern. In the `catch` block we can handle any errors that the method returns.

4. We package `playerName` (the single argument to this method) into a Dictionary, using our constant `kPlayerName` as the key, with which the name will be retrieved by the recipient of the message.

5. We send the Dictionary with the call to `updateApplicationContext` method of the `WCSession` class, and log to the console any errors that occur.

## Instantiate the class in AppDelegate

We want to make the first call to `PhoneConnectivityManager.sharedManager` (which is the call that will actually instantiate the shared instance of that class) early in the app's lifecycle, and the earliest place we can do that is when the `AppDelegate` class (created for us by Xcode's template) is itself instantiated. Add the following line of code to the `AppDelegate` class in the `AppDelegate.swift` file:

```
@UIApplicationMain
class AppDelegate: UIResponder, UIApplicationDelegate {

    var window: UIWindow?
    let connectivityManager = PhoneConnectivityManager.sharedManager

    ...
```

# Getting the user name

Now that we can send the player's name to the watch, all we need to do is get it from the player. We will simply add a **TextField** object (to input the name) and a **Button** object to the iPhone app's view (to trigger the send methods), and hook them up to some simple code in the `ViewController` class, which we shall write first.

## ViewController

Select `ViewController.swift` in the project navigator and add the following `IBOutlet` to the `ViewController` class:

```
@IBOutlet weak var playerNameTextFile: UITextField!
```

Then add the following methods to the class:

```
@IBAction func saveNameButtonTapped(sender: AnyObject) {
    savePlayerName() //1
}

func savePlayerName() {
    if playerNameTextField.text != "" { //2
PhoneConnectivityManager.sharedManager.sendNameToWatch
(playerNameTextField.text!) //3
    }
    "playerNameTextField.resignFirstResponder ()//4
}
```

1. The `saveNameButtonTapped` method, to which we will connect the button, is kept as slim as possible, with a single call to the `savePlayerName` method.
2. If the text input field is empty, we do nothing.
3. If a name has been entered, we call the `sendNameToWatch` method of the `PhoneConnectivityManager` class, which will take care of the rest of the message sending process.
4. To dismiss the keyboard that appeared when the user tapped on the `TextField` object, we call its `resignFirstResponder` method.

## Prepare the UI

In the `Main.storyboard` file, drag a **Button** object onto the interface, placing it in the center of the view using Interface Builder's layout guidelines.

Change its **Title** from **Button** to something more informative, such as **Send Name To Watch**.

We must now add **Layout Constraints** to the button, so that it will position itself correctly whatever the orientation and size of the view in which it appears. We can do this directly in Interface Builder:

*Control-drag* from the button onto the containing view at an angle of roughly 45 degrees, and release the mouse. You will be presented with the following contextual menu:

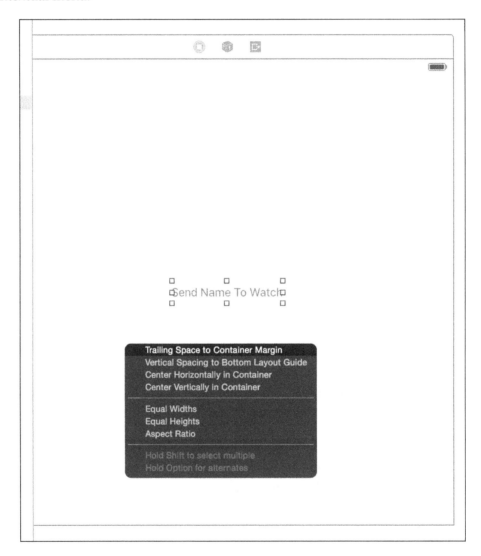

Follow these steps:

1.  Select **Center Horizontally In Container**, and then control-drag again to repeat the process, this time selecting **Center Vertically In Container**.

2.  Now drag a **TextField** object onto the view, and place it above the button, at a suitable distance (Xcode will help you once again with the layout guidelines).

3.  Drag the left and right edges of the text field to align with those of the button.

4.  Set its **Placeholder** text property to **Player Name**.

5.  Add a constraint as we did just now with the button, selecting Center **Horizontally In Container**.

6.  *Control-drag* from the center of the text field to the center of the button, and select **Vertical Spacing** from the contextual menu, to ensure that the text field will align itself at the same distance from the button regardless of the size and orientation of the view.

7.  The last constraint we will add will set the width of the text field. **Control-drag** from inside the text field to another point inside the text field, and select **Width** from the contextual menu.

## Hook up the UI

Using the same control-drag methods as earlier in *Chapter 3, C-Quence – A Memory Game* connect the `TextField` object in the Document Outline to the `@IBOutlet weak varplayerNameTextField: UITextField!`in the `ViewControllerclass`.

Connect the **Button** object to the `savePlayerName` method of the `ViewController` class.

# Creating the Watch Connectivity Manager

We must now return to the watch, where we will create a class, `WatchConnectivityManager`, that will use, to a large extent, the same code that we wrote for the `PhoneConnectivityManager` class.

1.  Select the `C-QuenceWatchKit` Extension folder in the project navigator, and then create a new Swift File (using *command-N*).

2.  Name it `WatchConnectivity.swift`, and make sure the **C-QuenceWatchKit Extension** `Target` is selected.

3.  Save it.

4. Import the `WatchConnectivity` framework and add the `WatchConnectivityManager` class to this file:

```
import Foundation
import WatchConnectivity

class WatchConnectivityManager: NSObject, WCSessionDelegate {

}
```

Just as we did with the `PhoneConnectivityManager` class, we will create a `sharedManager` method and declare the `init` method `private`, to ensure that only one instance of this class is created:

```
class WatchConnectivityManager: NSObject, WCSessionDelegate {

    static let sharedManager = WatchConnectivityManager()

    private override init() {
        super.init()
        if WCSession.isSupported(){
            let session = WCSession.defaultSession()
            session.delegate = self
            session.activateSession()
        }
    }
}

...
```

Because we have made `self` the delegate of the `WCSession`, we can provide a `didReceiveApplicationContext` method that will be called when the app receives, well, an `ApplicationContext` message. The method takes one argument, namely the `Dictionary` object we sent from the iPhone app:

```
    func session(session: WCSession,
didReceiveApplicationContextapplicationContext: [String : AnyObject])
{
        if let playerName = applicationContext[kPlayerName] as!
String? { //1
            print("received \(playerName) message") //2
                NSUserDefaults.standardUserDefaults().
setObject(playerName, forKey: kPlayerName) //3
        }
    }
```

1. We check if the `Dictionary` contains an entry with the key `kPlayerName` (you start to get why we made these constants?), and whether that entry is a **String** object (the method argument `Type` is `AnyObject`; other methods may be using this channel of communication).

2. If it is a `String`, we'll celebrate the fact by logging it to the console. You can probably remove this log later, once you're sure that everything is being called and behaving as it should.

3. We save the `String` to persistent storage, for use in the future, using `setObject` method of the `NSUserDefaults`. We save it here re-using the key `kPlayerName` (although any other key would do, so long as we know what it is when we need to retrieve the value from storage).

# Instantiate the class in ExtensionDelegate

On the watch side of things, it is even more important to create the shared instance of the `WatchConnectivityManager` class early in the app's lifecycle, since there may be messages waiting that were sent while the app was not running, so add the following code to the `ExtensionDelegate` class that was created by the Xcode template:

```
class ExtensionDelegate: NSObject, WKExtensionDelegate {

    let watchConnectivityManager = WatchConnectivityManager.
sharedManager

    . . .
```

# Modify the InterfaceController class

We are very close to being finished!

All we need to do now is add a method that retrieves the stored player name from `NSUserDefaults`. When the `resultLabel` method of the `InterfaceController` displays the player's score, it must first call this new method to find out what text to display.

Add the following method to the `InterfaceController` class:

```
    func textForResult(result: Int) -> String {
        if let playerName = NSUserDefaults.standardUserDefaults().
objectForKey(kPlayerName) as! String? {
            return "Hey \(playerName), you scored \(result)"
        } else {
            return "Not bad! You scored \(result)"
        }
    }
```

NSUserDefaults, as the name might imply, was originally intended to store small amounts of data like user preferences. It is the easiest way to store modest amounts of data that do not need to be saved in more specialized formats (for example, databases).

Once again, we use kPlayerName as a reliable way to check whether the NSUserDefaults contains a String with that key. We don't really need to check here if it is a String, because if it exists, it can't be anything else, but this check costs nothing, and may prevent trouble later on due to programmer mistakes.

 Program defensively! Assume every line of code you ever write is destined to become an international success that will be maintained by a small army of developers who are not as careful as you are.

Finally, we change the endGame method to call textForResult before setting the text of the resultLabel:

```
func endGame(result: Int) {
    gameboardGroup.setHidden(true)
    resultLabel.setHidden(false)

    let resultText = textForResult(result)
    resultLabel.setText(resultText)

    playButton.setHidden(false)
    playButton.setTitle("Play Again")
    gameLogic.clearGame()
}
```

# Run and test

Make sure that you have selected the WatchKit App Scheme and the correct iOS Simulator hardware:

Now hit **Run**. When you play the game, you will see no difference, because you have not yet set a player name.

So do this now. The watch app is running in Watch Simulator in the foreground, but that is not the case for the phone app:

In iOS Simulator, launch the **C-Quence** app. From here you can alter the player name and then lose a game on the watch, to see the personalized message appear in the result label.

Congratulations. The app is complete!

# Summary

This where we leave C-Quence, at least, as far as this book goes. It would be an excellent idea to go back over the code, right the way from the beginning of *Chapter 3, C-Quence – A Memory Game*, and experiment with tweaking the code here and there, with a view to better understanding what each line of code does, and how it might be varied.

And if it kills the app? No problem, that's what *command-Z* is for.

In this chapter, you have learned to:

- Iterate through the Run-test-fix process to catch bugs
- Adapt the originally planned code to real hardware conditions, You have
- Enabled phone-to-watch communication using the `WatchConnectivity` framework
- Use of constants to make that communication easier to code safely
- You have also learned to create shared manager singletons
- Use `NSUserDefaults` to persistently store data on both devices
- You have used Interface Builder's Layout Constraints to create a minimalistic interface on the phone that will correctly adapt to all screen sizes and orientations

In the next chapter, we will create a new app, and investigate, among others, the following topics: sharing larger data sets between devices, adding supplementary code to the phone; using Table objects in the interface; making use of the haptic interface—leveraging WatchKit's animations framework.

# 5
# On Q – A Productivity App

**On Q** will provide a little help to the user in remembering the bullet points of a presentation with no more than a glance at the watch, which will advance the prompts automatically without requiring interaction from the user. This app will also gather and store user data on the iPhone, and we will start to integrate the watch and phone apps to a greater extent than we have done up until now.

You will learn the following:

- Design and implement a UI that uses multiple interface controllers
- Use `xcassets` to store images
- Use those images to provide backgrounds to UI elements
- Add a menu that is summoned with Force Touch
- Add feedback from the Taptic Engine to your app
- Supply system callbacks to react to events outside the watch

## Download the project template

This project starts off a little differently. In order to focus on the watchOS code, we will begin with an Xcode project that already contains the code you will need for the iPhone app. This small amount of code is pretty standard stuff, and if you have done any iOS coding with Swift before, you will have no problems understanding what the code does.

You can download the project here:

```
https://github.com/codingTheHole/BuildingAppleWatchProjectsBook
```

Broadly speaking, the phone app simply enables the user to enter any number of prompts, and perform the standard re-order and delete tasks, which would be a troublesome task (to say the least) on the watch itself.

When you have downloaded and opened the project, you will find the complete iOS code, but for the WatchKit Extension and WatchKit app there is no more than the standard template files that you know already from our previous projects. By all means, have a look over the iPhone app source code. Anything that does not look familiar will become clear as we progress through this chapter.

# Plan the Watch app

Before we code, we plan. An hour's planning will generally save many hours' wasted coding. This is a good thing, and we've mentioned it before. We might even mention it again.

# Mission statement

We will create an app that will prompt the user with small snippets of text for use in say, a presentation or lecture, or any other scenario in which she might otherwise rely on cue cards or similar. Each prompt screen can be swiped to a second screen that shows more detailed, scrollable text. The prompts can be color-coded if desired, perhaps as an indication to progress a slide presentation, or hand out printed material and so on..

The prompts can be advanced manually by tapping forward and backward buttons, or automatically each time the user raises her wrist to glance at the watch screen. The succession of prompts can be stopped by using Force Touch to show a menu, from which the user can stop or continue the sequence.

The text of the prompts (and the detailed text that accompanies each one) is entered into the iPhone, along with the desired color preset, and when the user has edited the list to her satisfaction, she sends the complete list to the Watch, which will replace the old list with the new one and store it locally, so that the watchOS app can be used when the iOS app is not available.

# User story

As in the *Chapter 3, C-Quence – A Memory Game,* our first step to formalizing the app's requirements is to create a clear outline of what the user should experience when using the app.

# The Watch app

When the user launches the app, he is presented with a single **Start** button. On tapping this button, he is presented with a request to load prompts from the phone if there are none stored on the watch or he is presented with the first prompt if there are some loaded. Each prompt may be presented in one of the three colors, which are stored along with the text.

Each prompt screen contains **Forward** and **Backward** buttons, but lowering and raising of the watch will also advance the current prompt. To stop the prompts, the user uses Force Touch, a harder press on the watch touchscreen to access a menu that will enable him to return to the start button (or continue on if he changes his mind or activated the menu inadvertently).

If the current prompt is the first in the list, the **Backward** button will be grayed out and inactive. If it is the last prompt, then the **Forward** button will be grayed out inactive.

The user is kept informed by a label of how many prompts there are available, and which one is currently selected. This label is only visible after tapping the **Start** button.

From each prompt, the user will be able to swipe sideways to a screen that shows further text in a scrollable form, but which offers no navigation except to return to the screen from which he reached this one.

Let's reduce this to an outline of what the app will provide to the user:

## Launch view:

User taps Start button:

- Check data manager for prompts data

If no prompts are loaded:

- Show label requesting user to send them from phone

If prompts are loaded:

- Show prompts and navigation
- Hide Start button (and label, if showing)

## Prompts view:

Show (first) prompt:

- Get prompt from data manager
- Set `text` and `textColor`
- Enable **Forward** and **Backward** buttons as appropriate

If user swipes left:

- Show detailed text

If user taps Forward/Backward buttons:

- Show next prompt (as above)

User raises wrist to activate app:

- Show next prompt (as above)

User deploys Force Touch to display menu:

- Show menu

## Menu:

- Get user input

If player selects **Cancel:**

- Dismiss menu

If player selects **Stop**:

- Dismiss menu
- Hide prompts view
- Show **Start** button

Note that this list form is from the user's perspective and includes none of the behind-the-scenes action like connecting to the paired device.

# The iPhone app

On launching the app, the user will be able to navigate to a list of the prompts currently stored on the phone. From this list view he will be able tap an **Edit** button to re-order or delete prompts, as well as create a new one by tapping a + button. Tapping the + button will in turn launch a third view, which will allow the user to enter the prompt text as well as an optional detailed text, and choose one of the preset colors (default will be white, since the watch app will have a black background). User will be able to save this prompt or cancel; either way he is returned automatically to the list of prompts.

On navigating back to the launch screen, the user will be able to tap a button to send the new prompts list to the Watch, where it will replace the old contents if there are any.

All of this functionality is coded for you in the downloaded project file.

# App requirements

As we start to On Q:requisites" plan which classes we will require, we can now get a fair idea of the flow of the program, and therefore much of its functionality, from these outlines.

We will need to create a class that manages the app's data that is encapsulated away from direct interaction with the user, and just as with *C-Quence*, we will need a class that takes care of connectivity between the devices.

We will of course also add code to the files that Xcode creates for us as part of the WatchKit App template that we use to initiate the project, replace the InterfaceController file with something less generic, and create a second class of WKInterfaceController that will display the optional detailed text that may accompany a On Q:requisites prompt.

Note that the downloaded project template includes a `SharedConstants.swift` file that is a member of both the **On Q** and **On Q WatchKit Extension** targets, as pictured below (type *Command-Option-1* if the File Inspector is not visible):

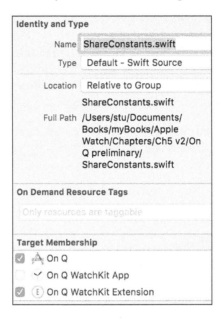

# Setting up the project

In this chapter, we are once again dealing with two apps in parallel, one for watchOS and one for iOS (which has been coded for you, but we will refer to it frequently). All but one of the files (`SharedConstants.swift`) will be used by one target (that is, one of the apps) or the other, and we must ensure that these targets are correctly selected for each file that we create within the project.

# Create the required watchOS classes

We will begin by creating the Swift files that we will need. Once this is done, you will have a good overview of the overall structure of the project and an idea of how the pre-coded iOS files relate to the watchOS app you will be writing.

In the project navigator, select the **On Q WatchKit Extension** group so that the files we create below will automatically select the WatchKit target in the file creation dialog (where we selected both the **On Q** and **On Q WatchKit Extension** targets above) and then perform the following steps:

1. Create a new Swift file, once again using *Command-N* **watchOS | Swift File**.

2. Name it `WatchConnectivity.swift`, and select the **On Q WatchKit Extension** target.

3. Hit **Create**.

4. Delete the `InterfaceController.swift` file.

Repeat steps 1-3 to create the following files:

- `WatchData.swift`
- `PromptsInterfaceController.swift`
- `DetailsInterfaceController.swift`

Note the close relationship between the Watch app and part of the iPhone app's file structure:

- `PhoneConnectivity.swift`
- `PhoneData.swift`
- `PromptsTableViewController.swift`
- `NewPromptViewController.swift`

Your project navigator should now look like this:

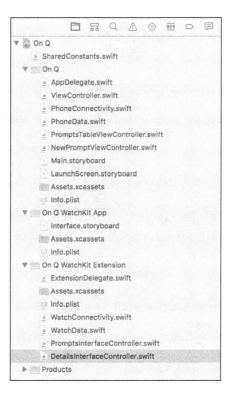

Even at this early stage of the project, we are able to see at a glance how our app will be structured and it would be perfectly reasonable to see this as the completed first phase of development. We have, after all, invested considerable time and energy in identifying what we want our app to do, how the user will experience her interaction with it, and what we as developers will need in order to prepare the ground for realizing the outline that we have created.

# Writing the code

So, time to get some code written. We will start with some code that should look very familiar.

# WatchConnectivity

Once again, we need to create a class that encapsulates all of the logic around communication with the outside world (or at least the part of the outside world that is squeezed into an iPhone).

## WatchConnectivityManager class

In the project navigator, select the WatchConnectivity.swift file. Delete the import Foundation line from the code, and replace it with the following code:

```
import WatchConnectivity

class WatchConnectivityManager: NSObject, WCSessionDelegate {

    static let sharedManager = WatchConnectivityManager()

    let dataManager = WatchDataManager.sharedManager

    private override init() {
    }

    func session(session: WCSession,
didReceiveApplicationContext applicationContext: [String :
AnyObject]) {
    }
}
```

Don't be alarmed by the error warning that Xcode shows about the dataManager; this is only happening because we haven't yet written the WatchDataManager class (we have only created the file into which we will write it). We'll do that next.

There is little else to be said here, you have seen all of this before.

# WatchData

Now we need to think about what we will require the `WatchDataManager` class to do for us.

The class will be a singleton, and will need a `sharedManager` class property, as we have seen in previous chapters.

One thing that is worth bearing in mind already is the fact that we have two different `WKInterfaceController` classes, one for prompts, one for detailed text, which are unable to talk directly to each other. We need somewhere to store the state of the app at any particular time, and since this state amounts to no more than a record of which data is available and which prompt is currently selected, the `WatchDataManager` class seems a suitable place. So this class will need methods to provide the contents of whichever prompt is selected, as well as a number of methods that will provide other classes with information about the `prompts array` that we will implement later.

## WatchDataManager class

Select `WatchData.swift` in the project navigator, and replace `import Foundation` with the following code:

```
import WatchKit

class WatchDataManager {

    static let sharedManager = WatchDataManager()

private init(){ //1
    }
    func nextPrompt() -> Prompt? { //2
            return nil
    }
    func previousPrompt() -> Prompt? {
            return nil
    }
    func isFirstPrompt() -> Bool {//3
        return false
    }
    func isLastPrompt() -> Bool {
        return false
    }
    func lastSelectedPrompt() -> Prompt? { //4
            return nil
```

```
    }
    func lastSelectedPromptIndex() -> Int { //5
        return 0
    }
    func promptsCount() -> Int { //6
        return 0
    }
    func updatePrompts(prompts: [Prompt]) { //7
    }
    func reset() { //8
    }
}
```

The comments in the code are as follows:

1. In the `init` method (which once again, is private to force other classes to use the `sharedManager` property to access the single instance of this class), we will eventually retrieve any stored `Prompt` data, but we will at first stub this method with some fake data until we have enable the phone app to get the real stuff from the user (see method implementation below).

2. The next prompt or previous prompt will be called from the `InterfaceController` class. Both the `nextPrompt` and `previousPrompt` need to be able to return either a `Prompt` if there is one, or `nil` if there is not, and so their return value in the optional `Prompt?`, which may return `nil`.

3. Since the **Forward** and **Backward** buttons will be made inactive at times, the interface controllers will need to ask the `sharedManager` if the current prompt is the first or last.

4. The `DetailsInterfaceController` class will need to know which prompt the `InterfaceController` is currently displaying.

5. The label informing the user of where she is in the prompts list will need the current prompt's position in that list.

6. We will not expose the `Array` of `Prompt` objects that the class will maintain to other classes directly, but enable them to get the number of items in the array. This method will be called by the label that shows the user her position in the sequence of prompts.

7. This method will be called whenever the `WatchConnectivityManager` receives data from the phone app.

8. When the user stops a session from the Force Touch menu, the `WatchConnectivityManager` will need to do a little clearing up.

# Interface Controllers

For the first time, we have more than one type of interface controller, that is, two separate subclasses of `WKInterfaceController`.

## PromptsInterfaceController

Replace the `import Foundation` statement with the following:

```
import WatchKit
```

Then add the following code, to create the class:

```
class PromptsInterfaceController: WKInterfaceController {
}
```

Here, we create the `PromptsInterfaceController` class and declare it to be a subclass of `WKInterfaceController`.

We will need to override two of the `WKInterfaceController` methods that are called by the system as the app goes in and out of the active state in order to automatically advance the current prompt when the user raises her wrist. Add the following methods inside the class's curly braces:

```
override func willActivate() {
    }

override func willDisappear() {
    }
```

Lastly, we add the custom methods that will contain the rest of the class's logic:

```
func start() {
    }
func nextPrompt() {
    }
func previousPrompt() {
    }
func updateUIForPrompt(newPrompt: Prompt) {
    }
func updateButtonStates() {
    }
```

## DetailsInterfaceController

Firstly, replace the `import Foundation` statement with the following:

```
import WatchKit
```

Then add the following code to create the class:

```
class DetailsInterfaceController: WKInterfaceController {
}
```

With our interface controllers classes declared as subclasses of `WKInterfaceController`, we will now be able to hook up the Interface Builder storyboard to the code.

# The Storyboard

We will be adding images to this project in order to create a more engaging interface, and so it is worth spending a little while looking at what this involves in relation to building a WatchKit app.

## Preparing the images

Now, a detailed look into graphics is way beyond the scope of this book, so here's what you really must know about creating images for the Apple Watch:

- The 38mm watch has a screen size of 272 x 340 pixels, and the 42mm watch has a screen size of 312 x 390 pixels. There is no point in using 1024 x 1024 pixel images, it's just a waste of space, processor time and bandwidth. Keep 'em small.

- Apple recommends the `.png` image format, and we will stick with that, where possible. If a photo editing app needed to display `.jpeg` files, we could do that, but where we create the resources ourselves, we might as well create them as `.png`.

- All images are for Retina displays (there are no non-Retina Apple Watches) so the filenames require the `@2x` at the end of the file name (but before the `.png` suffix), otherwise they will not load.

The images used in this chapter are, of course, available to download from the book's github repository:

```
https://github.com/codingTheHole/BuildingAppleWatchProjectsBook
```

# Using xcassets

Let's assume you have downloaded the `ch5RedBg@2x.png` file from the website.

In the **On Q WatchKit App** group in the project navigator, select the `Asset.xcassets` group (be sure that it's not one of the `Assets.xcassets` groups from the other two targets).

Let's say that again. It must be the right `Asset.xcassets` group, the one that belongs to the WatchKit App target. If you find your graphics not showing up in the app, check this first. This is the voice of weary experience; this is such an easy mistake to make.

From the Finder, drag the file into the part of the project window that contains the **AppIcon** image set:

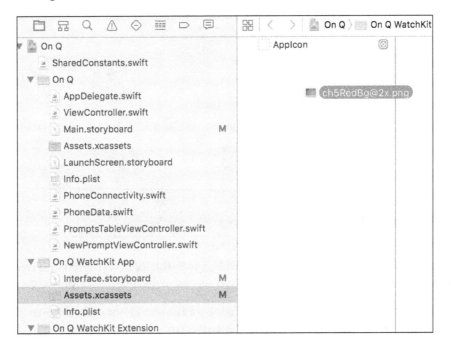

Xcode will then automatically create an image set for you and store a copy of the image file in your project folder, as shown below:

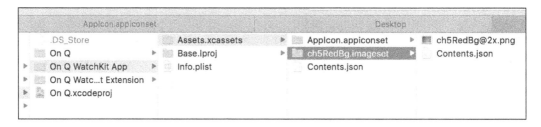

In the Attributes Inspector pane under **Devices** (*Command-Option-4*), select **watchOS | Apple Watch** and deselect **All | Universal** (this image will only be used on the watch) as illustrated here:

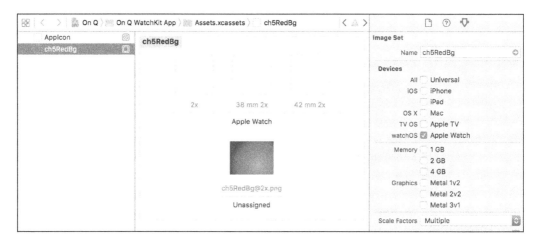

Now drag the image from the **Unassigned** box into the **2x** box (there is no need here for us to create separate versions for the two screen sizes, even though we can).

Organizing your images into image sets using xcassets is not only the way recommended by Apple, it's also by far the easiest and safest.

# Designing the user interface

Now that we have our image assets prepared and integrated into the project, we can start work on the UI.

In the **On Q WatchKit App** group in the project navigator, select **Interface.storyboard**.

# PromptsInterfaceController

We now need to associate the **InterfaceController** with our `PromptsViewController` class.

In Interface Builder, select the **InterfaceController** and use the Identity Inspector pane (*Command-Option-3*) to set its **Custom Class | Class** property to **PromptsViewController**, as shown:

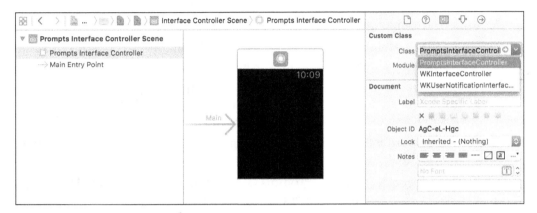

Now when you have the Assistant Editor showing (*Command-Option-Return*), it will show the `PromptsViewController` class, where you can drag across the IBActions and Outlets that are necessary.

When you wish to hide the Assistant Editor, use the *Command-Return* keyboard shortcut.

## Prompts Group

If you are still looking at the Identity Inspector pane, change to the Attributes Inspector (*Command-Option-4*).

We will create the group of UI elements first that are hidden at app launch, then we can mark them as hidden and they will be out of our way while we concentrate on the start group.

Using the techniques that we have covered in previous chapters, perform the following steps:

1. Drag a **Group** object onto the interface.
2. Set its **Layout** property to **Vertical**, and its **Height** property to **Relative To Container**.

3. *Control-drag* from the Group into the Assistant Inspector to add an **Outlet** which should be named **promptsGroup**. In the Organizer pane on the left of the IB window, you'll see that Xcode has conveniently renamed the Group **Prompts Group** to match. Neat touch.

## PromptLabelGroup

1. Drag a **Group** object into the **Prompts Group**, and rename it **Prompt Label Group** in the Organizer (we don't create an Outlet here, so we have to rename it manually).

2. Set its **Size | Height** property to **Relative To Container**, and change its value to 0.55 in the text field immediately below that.

3. Drag a **Label** object into the **Prompt Label Group**, setting its number of **Lines** to **0** (this will allow the label to resize to accommodate as many lines as necessary), and its **Horizontal** and **Vertical Alignment** properties to **Center**.

4. *Control-drag* from the label to the Assistant Editor to add an Outlet and name it **promptLabel**.

5. Drag a Label object into the Prompts Group (not the Prompts Label Group!) and set its Horizontal Alignment to Right, and its Vertical Alignment to Bottom.

6. *Control-drag* from the label to the Assistant Editor to add an Outlet and name it **currentPromptNumberLabel**.

## Buttons Group

Now let's move on to the forward and backward buttons, which we will place inside a group of their own:

1. Drag a **Group** object into the **Prompts Group**, underneath the **Current Prompt Number Label**, set its **Vertical Alignment** property to **Bottom**, and set its **Height to Relative To Container** with a value of 0.3. Manually name it **Buttons Group**.

2. Drag a **Button** object into the **Buttons Group**, set its **Title** to **<** and its **Vertical Alignment** to **Center**.

3. Set its **Font size** to 28, its **Width** to **Relative To Container** with a value of 0.5, and its **Background Color** to **Clear Color**.

4. Copy and paste this button within the **Buttons Group**, and change the copied button's **Title** to **>** and its **Horizontal Alignment** to **Right**.

5. *Control-drag* from these buttons into the code to add Outlets named **backwardButton** and **forwardButton**.

6. *Control-drag* from the **Backward** button into the code, and select **Action** as pictured below:

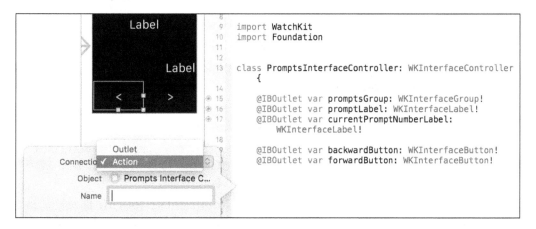

7. Enter the name **backwardButtonTapped** and click **Connect** to add an IBAction to the class's code.

8. Do the same for the other button, naming the Action **forwardButtonTapped**.

Your Prompts Group should now look like this:

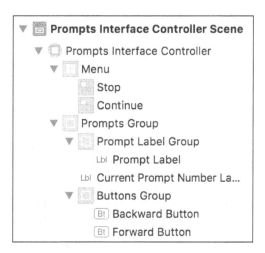

We have finished the **Prompts Group**, so select it, and set it to **Hidden** by ticking the appropriate box in the Attributes Inspector.

# Start Group

Now with the **Prompts Group** out of the way, we can create the **Start Group**.

1.  Drag a **Group** object onto the (now seemingly empty) interface.

2.  Set its **Insets** property to **Custom**, and set all four inset values to **2**.

3.  Select the button image that you added to the `Images.xcassets` file in the **Background** drop-down menu.

4.  Set the **Radius** to **Custom** with a value of **8**, the **VerticalAlignment** to **Center**, and the **Height** to **Relative To Container**, with a value of **0.3**.

5.  *Control-drag* into the `PromptsInterfaceController` class in the Assistant Editor to create an outlet, and name it **startGroup**. This name will be reflected in the organizer, as we have seen before.

6.  Now drag a **Button** object into the **Start Group**, and set its **Title** to **Start Prompting** (or something similar, it's completely cosmetic), set its **Background Color** to **Black Color**, and its **Height** to **Relative To Container**.

7.  *Control-drag* from this button into the code, below the `forwardButton` Outlet, select**Action** from the contextual pop-up and name the action `startButtonTapped`.

# Other UI elements

We now need to add a label that will request that the user load data from the phone when none has been stored on the device, as well as a **Menu** object with which the user can stop the prompts:

1.  Drag a **Label** object onto the interface below the **Start Group** (don't worry that it appears above the **Start Group** when you release the mouse, we'll fix that in a second).

2.  Set its Text to Please send Prompts from your iPhone, its Text Alignment to be centered, and its number of Lines to 0.

3.  Set its **Vertical Alignment** to **Center** (now it appears below the **Start Group** as it should) and its **Width** to **Relative To Container**.

4.  *Control-drag* into the code and create an Outlet named **loadPromptsLabel.**

5.  Now set it to **Hidden** (we don't need to see it on launch), and the **Start Group** will once again be in the center.

6.  Drag a **Menu** object onto the interface (it doesn't matter where, you won't even be able to see it except in the Organizer pane on the left).

7.  Set its number of **Items** to **2**.

8. Now you can select each **Menu Item** individually in the Organizer (after clicking the disclosure triangle). Set the **Title** of the first item to **Stop** and select the **Pause** image from the **Image** list (or you could try creating your own!).

9. *Control-drag* from the **Stop** item into the code to create an Action, and name it **menuStopButtonTapped**.

10. Set the title of the second item to **Continue** and set its image to **Play**.

Your Organizer should now look like this:

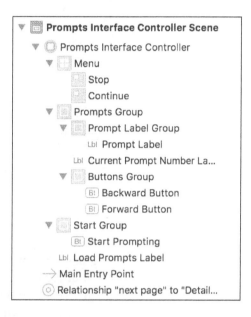

# DetailsInterfaceController

Now we will create the interface of the `DetailsInterfaceController`, which will simply display the detailed text that may accompany each prompt.

1. Drag an **InterfaceController** object into the Storyboard, next to the **Prompts Interface Controller**.

2. *Control-drag* from the **Prompts Interface Controller** (but NOT the **Start Group**) into the new Interface Controller, and select **next page** from the **Relationship Segue** pop-up that appears, as pictured here:

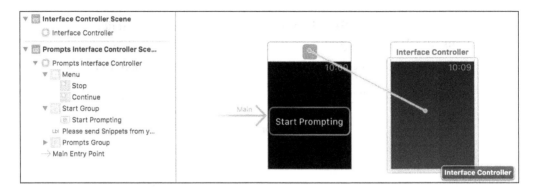

 **Segue** is the name that Apple has borrowed from the world of music (where it means a smooth and continuous transition from one song to the next) for its method of handling transitions from one screen to another. You can either add segues by writing code yourself, or simply by using the Interface Builder. We'll be seeing more of Segues later on.

3. That's all you have to do to add a view which you can swipe from the **Prompts Interface Controller**! Give it a go now — select the **On Q WatchKit App** scheme next to the **Run** button, hit **Run** and you'll see the app launch and show; in addition to our fine looking Start button there are two dots at the bottom of the screen, indicating how many pages are available and which one is currently visible. Now you can swipe between these screens.

4. Back in Xcode, we need to associate this controller with our `DetailsInterfaceController` class, which we do again by selecting that class in the Identity Inspector (*Command-Option-3*), illustrated below:

5. Now drag a **Label** object onto the interface, set its **Text Alignment** to be centered, its number of **Lines** to **0**, and its **Width** to **Relative To Container**.

6. *Control-drag* from this label to the source code to create an Outlet and name it **detailsLabel**.

Interface controllers don't come much simpler than that. But it really is all we need for this particular app.

The rest of the magic is in the code!

# Implementing the methods

The first two classes we will complete, ExtensionDelegate and WatchConnectivityManager, will be very familiar to you since they will hardly differ from the implementations we have seen in the *C-Quence* project.

## ExtensionDelegate

As we did previously, we need to initiate our WatchConnectivityManager as early as possible in the app's lifecycle, so we do this in the ExtensionDelegate class. Add the following code to the ExtensionDelegate.swift file generated automatically by the Xcode Template:

```
class ExtensionDelegate: NSObject, WKExtensionDelegate {
        let watchConnectivityManager = WatchConnectivityManager.
sharedManager
. . .
```

## WatchConnectivityManager

The code for the WatchConnectivityManager is almost identical to what we have seen before and, apart from noting the use of the kPromptsKey and Prompttypealias, needs no further comment here:

```
import WatchConnectivity
class WatchConnectivityManager: NSObject, WCSessionDelegate {
    static let sharedManager = WatchConnectivityManager()
    let dataManager = WatchDataManager.sharedManager
    private override init() {
        super.init()
        if WCSession.isSupported(){
            let session = WCSession.defaultSession()
            session.delegate = self
            session.activateSession()
        }
```

```
    }
    func session(session: WCSession,
didReceiveApplicationContextapplicationContext: [String : AnyObject])
{ // non main queue
        if let prompts = applicationContext[kPromptsKey] as! [Prompt]?
{
            WatchDataManager.sharedManager.updatePrompts(prompts)
        }
    }
}
```

# WatchDataManager

When we implement the code for WatchDataManager, we will use some dummy prompts data for testing purposes, but we will replace the stubbed return values of these methods with proper code.

Add the following code to the WatchDataManager class as indicated:

```
import WatchKit
let kPromptsArchive = "promptsArchive" //1
class WatchDataManager {
    static let sharedManager = WatchDataManager()
    static let DocumentsDirectory = NSFileManager().URLsForDirectory
(.DocumentationDirectory, inDomains: .UserDomainMask).first!
static let PromptsArchive =
DocumentsDirectory.URLByAppendingPathComponent(kPromptsArchive) //2

    private var currentPromptIndex = -1 //3
    private var prompts: [Prompt] = []

    private init(){ //4
        prompts = [
            [kPromptText: "Example Prompt",
                kDetailsText: "Example Details",
                kColor: kWhite],
            [kPromptText: "Second Prompt",
                kDetailsText: "More Details",
                kColor: kRed],
            [kPromptText: "Third Prompt",
                kDetailsText: "Even More Details",
                kColor: kGreen]
            ]
    }
    func nextPrompt() -> Prompt? { //5
```

```
            if currentPromptIndex >= prompts.count - 1 {
                return nil
            } else {
                currentPromptIndex += 1
                return prompts[currentPromptIndex]
            }
        }
        func previousPrompt() -> Prompt? {
            if currentPromptIndex <= 0 {
                return nil
            } else {
                currentPromptIndex -= 1
                return prompts[currentPromptIndex]
            }
        }
        func isFirstPrompt() -> Bool { //6
            returncurrentPromptIndex <= 0
        }
        func isLastPrompt() -> Bool {
            return currentPromptIndex >= prompts.count - 1
        }
        func lastSelectedPrompt() -> Prompt? {
            if currentPromptIndex >= 0 && currentPromptIndex < prompts.count {
                return prompts[currentPromptIndex]
            } else {
                return nil
            }
        }
        func lastSelectedPromptIndex() -> Int {
            return currentPromptIndex
        }
        func promptsCount() -> Int {
            return prompts.count
        }

        func updatePrompts(prompts: [Prompt]){ //7
            NSUserDefaults.standardUserDefaults().setObject(prompts,
    forKey: kPromptsKey)
            self.prompts = prompts
            WKInterfaceDevice.currentDevice().playHaptic(.Success)
        }

        func reset() { //8

            currentPromptIndex = -1
        }
    }
```

The comments in the code are as follows:

1.  We create a string constant which will be used as the key to store the data we receive from the phone (see `updatePrompts`).

2.  We use `NSFileManager` to get the URL of our app's documents directory (provided by the OS), and append the name of our stored data key to create the URL to our persistent data.

3.  The `currentPromptIndex` is set to `-1`, indicating that no prompt is selected. In this way, the first increment of that variable gives us a value of 0, which is the index of the first prompt.

4.  We create some stub data. Later we will replace this code to read data from disk.

5.  The next two methods maintain a record of which prompt from the data is selected, returning the current prompt if there is one, or `nil` if none is selected. The return type is therefore the optional `Prompt?` instead of `Prompt`.

6.  The next few methods will be called by the interface controllers to retrieve information about the current prompt.

7.  Here we use `NSUserDefaults` to store our `prompts` data to disk, and trigger the `.Success` feedback from the Taptic Engine.

# Interface Controllers

Now we can add the code for the `InterfaceController` classes.

## PromptsInterfaceController

Add the following property declarations below the `IBOutlets` of the `PromptsInterfaceController` class:

```
let dataManager = WatchDataManager.sharedManager //1
var expectedDisappear: Bool = false //2
var running: Bool = false
```

The comments in the code are as follows:

1.  The `dataManager` will be the source of all the app's data, the interface controller needs no direct knowledge of where that data comes from, or how it is obtained.

2. The `expectedDisappear` and `running` properties will enable the class to distinguish between changing between views within the app, and the view appearing due to the app having been deactivated (which we use to automatically advance the prompt). This will become clearer as we fill in the rest of the methods.

Next, we'll add the code to the `IBActions` that are called by the UI:

```
import WatchKit
class PromptsInterfaceController: WKInterfaceController {

    . . .

    @IBAction func startButtonTapped() { //1
        start()
    }
    @IBAction func forwardButtonTapped() {
        nextPrompt()
    }
    @IBAction func backwardButtonTapped() {
        previousPrompt()
    }
    @IBAction func menuStopButtonTapped() { //2
        running = false
        promptsGroup.setHidden(true)
        startGroup.setHidden(false)
    }
```

The comments in the code are as follows:

1. As previously, we keep the `IBActions` as slim as possible, they do no more than call a single method that will provide the logic.

2. Tapping the menu's **Stop** item sets the app back to the state it would be in immediately after launch. Where is the **Continue** method? We don't need one – the menu closes after the user makes a selection, and there is no more to be done – the app just continues as if the menu had never been activated.

Next we will implement the controller's methods that are called by the OS by adding the following methods to the class:

```
override func willDisappear() { //3
        expectedDisappear = true
    }
```

```
override func willActivate() { //4
    if !expectedDisappear && running {
        nextPrompt()
    }
    expectedDisappear = false
}
```

1.  If we leave the view by user navigation to another view, we store this in the `expectedDisappear` property.

2.  When the controller enters the active state, we check if the user navigated away from it, or whether the app has been deactivated (by lowering her wrist, for example). If the prompts are `running` and the deactivation of the controller was not through navigation, then we will call the `nextPrompt` method (that we will write shortly), just as if the user had tapped the **Forward** button.

```
func start() {
    if dataManager.promptsCount() == 0 { // 5
        loadPromptsLabel.setHidden(false)
    } else {
        buttonGroup.setHidden(true) // 6
        loadPromptsLabel.setHidden(true)
        runningUIGroup.setHidden(false)
        dataManager.reset() // 7
        running = true
        nextPrompt()
    }
}
```

3.  If the user has not yet loaded any prompts data from the phone, we show the text **Label** requesting her to do so.

4.  If there is data available, we hide the `startGroup` and `loadPromptsLabel`, and show the `promptsGroup` of UI elements.

5.  We reset the `dataManager` to its starting state (whether we are starting for the first time or we have been running before and stopped from the menu), and set the `running` flag to true.

```
func nextPrompt() {
    if let newPrompt = dataManager.nextPrompt(){ //8
        updateUIForPrompt(newPrompt)
    }
}
```

```
func previousPrompt() {
    if let newPrompt = dataManager.previousPrompt(){
        updateUIForPrompt(newPrompt)
    }
}
```

6. We check if a nextPrompt (or previousPrompt) exists, and if it does we update the interface with the new data.

```
func updateUIForPrompt(newPrompt: Prompt) {
    if let promptText = newPrompt[kPromptText] { //9
        promptLabel.setText(promptText)
    }
    if let promptColor = newPrompt[kColor] {
        promptLabel.setTextColor(globalColors[promptColor])
    }
    updateButtonStates() // 10
}

func updateButtonStates() {
    forwardButton.setEnabled(!dataManager.isLastPrompt())
    backwardButton.setEnabled(!dataManager.isFirstPrompt())

    if dataManager.lastSelectedPromptIndex() >= 0 {
        currentPromptNumberLabel.setHidden(false)
        currentPromptNumberLabel.setText(
"\(dataManager.lastSelectedPromptIndex() + 1) of
\(dataManager.promptsCount()) ")
    } else {
        currentPromptNumberLabel.setHidden(true)
    }
}
```

7. We check that there is data available and if there is, we set the promptLabel's properties accordingly.

8. We then call updateButtonStates which polls the dataManager for information with which we can set the visibility of the currentPromptNumberLabel and buttons as appropriate.

# DetailsInterfaceController

The implementation of the `DetailsInterfaceController` class is very simple.
Add the following:

```
import WatchKit

class DetailsInterfaceController: WKInterfaceController {

    @IBOutlet var detailsLabel: WKInterfaceLabel!

    let dataManager = WatchDataManager.sharedManager //1

    override func willActivate() {
        if let newPrompt = dataManager.lastSelectedPrompt(){ //2
detailsLabel.setText(newPrompt[kDetailsText])
        } else {
            detailsLabel.setText("Please load Prompts from your
Phone")
        }
    }

    override func willDisappear() {
        detailsLabel.setText("") //3
    }
}
```

The comments in the code are as follows:

1. We will need a reference to the `WatchDataManager` shared instance
2. We check if there is a selected prompt. If there is, we set the `detailsLabel` text accordingly, if not we ask the user to load them.
3. Before the view disappears, we clear the `detailsLabel` text, so that the screen is empty next time we see it.

# Build and run

You can now run the app and you will be able to use it in all its functionality.
The prompts will be less than thrilling, but we'll be getting to that very soon.

# Test and Tweak

If you have followed the example code correctly, you should not have any issues
using the app. That's not to say that the design and the user experience could not be
improved on, far from it, but what is there should be working fine. If not, check the
code carefully.

# Testing with the iPhone and real data

We are now ready to test entering user data into the iOS app (which should work without issues—you downloaded it from the website, after all) and to test the transfer of that data to the watchOS app.

If you have already run the iPhone app and tested what the buttons do (we're developers, we're naturally curious, OF COURSE you've played around with it), then you will already have sent some default data to the Watch. Let's get rid of that first by deleting the app from the iOS Simulator.

Now we're all starting from the same point.

# Unstub the WatchDataManagerinit method

Although we have stored prompts data to `NSUserDefaults` in the `updatePrompts` method, we still have a `dataManager` that returns stubbed data, so let's fix that now. Replace the `init` of the current `WatchDataManager` implementation with the following code:

```
private init(){
    if let storedPrompts = NSUserDefaults.standardUserDefaults().
objectForKey(kPromptsKey) as! [Prompt]? {
            self.prompts = storedPrompts
        }
    }
```

Now, finally, we can read in the stored data from disk if it exists. We retrieve the `Dictionary` of `Prompt`objects, which is stored with the `kPromptsKey`, and copy its value into the `prompts` property of the `dataManager` class, where it can be made available to the rest of the app.

 If you test the app on a real Apple Watch with a real iPhone, you'll need nerves of steel the first time it runs, and the patience of a saint on subsequent launches: A quick test just now took about four minutes before anything launched on the watch. You might find it expedient to do much of your testing using the simulators!

Of course, the first time we run the app after changing this code (and deleting the iOS app), we will see no change. We will simply see our label that requests we send some data from the iPhone.

To get some actual user defined prompts we must launch the iOS app on the iPhone and send across some data. Be careful here that you are launching the watch app from Xcode using the **On Q WatchKit App** scheme, but launch the iOS app from the simulator.

Now when you hit **Send Prompts** in the phone app and start prompting in the watch app, the default data from the iOS code ("Example Prompt") will be available on the phone. And of course, if you use the phone app to enter some slightly more colorful data (both metaphorically and literally), you'll be able to send that for use on the Watch.

If you are testing on the Watch Simulator, you can use *Command-L* to simulate the user lowering his wrist. If you are testing on an actual wrist, you can, well, lower your wrist. Now type *Command-L* again (or raise your wrist) and the app will automatically advance to the next prompt (assuming there is one).

Each time you tap **Send Prompts** on the iPhone, the old data will be overwritten on the watch, so go ahead and try some longer lists of prompts. Try a long text in the Details page, and you will see that `WKInterfaceLabel` gives you all the scrolling functionality for free, just so long as you have the **Height** property set to **Size To Fit Content**. If you are testing on a real watch, you'll also notice the Taptic feedback you get when the Watch wakes to discover it has received new data.

# Challenges for further growth

Time to jam around with the code, since an essential part of becoming familiar with any platform is building up a wealth of experience of what is possible, what is worth repeating and what doesn't work at all.

- Try changing the appearance of the app in more fundamental ways than simple changes of color (critical though that can be). Is there something you can do to make those labels more engaging? Can you fit the app's appearance to some (possibly invented) corporate image? Could the layout be improved?

- Can we add a button to the launch screen that will disable the auto-advance feature? We'd need to store that preference too.

- How would the user experience improve if we added a little animation magic to the transition from the launch screen's starting UI and its prompting UI? Any other candidates for animation?

- If you already have more than rudimentary experience with developing for the iPhone, you might wish to add the ability to store and recall multiple Prompt sets on the phone.

# Summary

In this chapter, you have learned to add navigation to separate screen pages using IB Segues, and you have created separate subclasses of `WKInterfaceController` to manage those pages. You have used Xcode's `xcassets` to add images to your project and learned how to use those images in your app's UI, and you have used the Apple Watch's Taptic Engine to add Haptic feedback, thus adding another layer of communication with the user.

In *Chapter 6*, *Watching the Weather* we will be looking at using `NSURLSession` to get the Watch linked to the internet, using other data transfer technologies when transferring data back to the phone from the watch, and we will also add a whole new source of interaction with your app, the so-called **Glance** screen.

# 6
# Watching the Weather

In this chapter we will expand the scope of our development skills, and begin to leverage the vast resources of the Internet. This will give us an opportunity to take a look at using open source data, how to get that data from the web using NSURLSession and how to handle the JSON, one of the web's most common data formats, and the one in which we will receive the information that we are after.

In addition to that, you will learn to present selected parts of the data in a table, as well as presenting selected data in the form of a **Glance**, which is available to the user even when the main part of your watch app is not running.

And all of this with no help from the iPhone! We will be connecting to the largest source of information in history, using the smallest smart device ever to have reached the mobile tech scene.

If this sounds a little daunting, rest assured that we will be focusing on the simplest methods that are available, using a very minimal user interface, in an app with a very restricted set of features. This chapter is much more about how to get and use data than how to present it (important though that is). Indeed, you will see that the *Weather Watch* app involves fewer lines of code than in *Chapter 5, On Q – A Productivity App*.

By the end of this chapter you will have deepened your knowledge and experience of watchOS programming to a very significant extent, having covered the following topics:

- Data in JSON format
- Open source data on the net
- Apple's App Transport Security
- Passing closures as arguments to functions

- Showing user alerts
- Using tables

# Adding a Glance to an app

So, there is much to be done. Let's plan out first what we are going to need.

## Plan the App

As we have done before, we will first get a clear idea of what we want the app to do before we write any code.

## Mission Statement

We will create an app that will download current weather data for a number of cities around the world, using data from an open source provider. The data will be stored on the device, and only refreshed after a specified period has elapsed, to keep data traffic to a minimum. The rationale here is that weather conditions don't change from one second to the next (except in Auckland), and so displaying data that has been previously downloaded makes more sense up to a certain point.

The data will be displayed in the form of a brief description next to the name of each city, presented in a table. Tapping on one of the table's rows will open a screen that will contain more details of the selected city, such as wind-speed, temperature, and the like.

One city will be singled out for special status, and the current weather summary for this city will be available as a Glance by swiping up from the watch face when the app is not running in the foreground.

## User Story

The user will open the app and be offered the opportunity to view the weather, on a screen dedicated exclusively to presenting that information (that is, no buttons or any other UI elements). On this dedicated screen, the user will be able to scroll through the rows of a table, each of which displays the name of a city and a brief textual summary of its current weather conditions. The user can choose to return to the launch screen and update the weather, or may tap a table row to reach a third screen that will contain additional information, as well as display an icon representing the weather conditions of that city.

We can sketch this out into a rough flow diagram as pictured here:

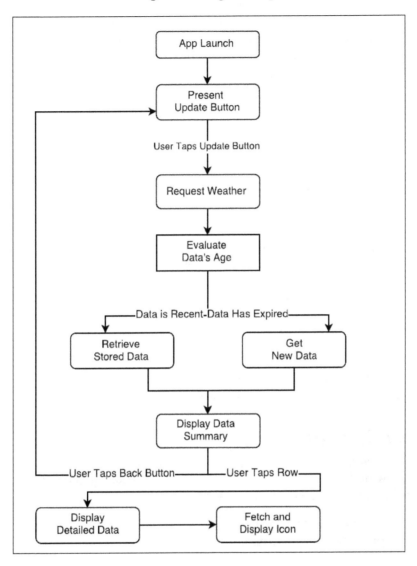

# Setting up the project

Set up a new project using the **iOS App with WatchKit App** template that we have used before, but this time include the **Glance Scene** (and deselect **Notification Scene** as previously), as pictured here:

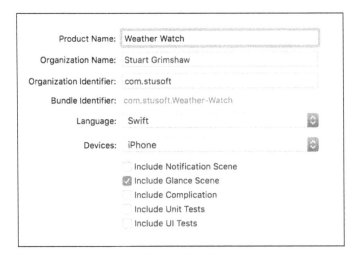

# Requirements

Using the flow diagram to provide some (very welcome) visual support, we can now start to think about our requirements in more concrete terms.

## Weather data structure

Since we'll be passing a number of weather summaries to the table view that displays the initial view of the data, it makes sense to be able to package the data necessary for each individual summary into some sort of structure.

Create a new Swift file, call it `WeatherData.swift`, and select both the WatchKit app and the iOS app targets, as pictured here:

Why do we select the phone app target when we are writing an app that will function exclusively on the watch? Well, this is simply some prudent future-proofing of the app. It seems likely that we will expand the app at some point in the future to provide the same data on the phone, and if we do so we will want to use the same data structures.

We will store the data in `struct`.

Paste the following code into `WeatherData.swift`, after the `import` statement:

```swift
struct WeatherSummary { //1
    let cityName: String
    let summary: String
    init(cityName: String, summary: String) {
        self.cityName = cityName
        self.summary = summary
    }
}

typealias WeatherData = [WeatherSummary] //2
```

The comments in the code are as follows:

1. We define `WeatherSummary` to be a Swift `struct` to contain the data, which has two properties; a string to hold the name of the city, and one to hold a summary of the weather conditions.

2. We define a `typealias` with which we can refer an array of `WeatherSummary` objects. This keeps the code elsewhere in the app easier to read.

That's it for the `WeatherData.swift` file.

# Getting the data

So how, and from where, do we get the data we need from the web? There are, in fact, very many sources that will provide us with data for free, and not just weather data. Searching through Google open data just now yielded 728 million results, which should be enough to keep most curious developers amused for some time.

# Welcome to openweathermap.org

From the large number of open weather data sources, we will select http://
openweathermap.org, which provides a large quantity of data for free to registered
users. Go to the URL above and perform the following steps:

1. Create a new account, which is free, requiring only an email address.

2. Copy the API Key that will be generated once you have created an account.

You're done! Now you have an API key that you will include when you request data
from the server, confirming that you are entitled to access that data. The key will
look something like this: **78c75588dfe58276a694af9c660edxxx**

With less form-filling than it takes to create a Google account you now have access to
extensive weather data, covering the entire planet. What an amazing thing software
development is.

Using http://openweathermap.org/ city ID's in our data requests, we can specify
the cities whose data we wish to display to the user. The code in this book uses the
ID's of a few cities dear to my heart, but you can change those to any cities you wish
by referring to the city ID's listed here:

http://openweathermap.org/help/city_list.txt

# Introducing JSON

Now, there are a number of formats in which you can download data from http://
openweathermap.org/, and we will be using JSON, which has become one of the
most popular formats due to its ease of use and lightweight structure. JSON stands
for JavaScript Object Notation, but you don't need any JavaScript knowledge to use
it, it just turned out to be a really convenient format.

JSON data is text. You can read it. You can print it to the console. And it is really
easy to use. It is also very easy to convert into Swift Dictionary objects using
NSJSONSerialization methods that are available in WatchKit. And once you
have done that, it is just a matter of parsing those dictionaries.

## JSON data structure 101

JSON data contains arrays and dictionaries, much the same as you know from iOS,
each of which contains further arrays and dictionaries, or strings, numbers and
Boolean values, collectively referred to as objects.

One of the first things you need to do when you have obtained JSON data and converted it to a `Dictionary`, is print that dictionary using `print(yourDictionary)`. You can then see, in the console, all the data that is available, and the structure in which it is stored. With a little familiarity with such data, you will soon be able to quickly inspect the dictionaries and arrays, in order to find the information you need.

## Making the data more readable

With deeply nested arrays of dictionaries that contain dictionaries of arrays and yet more dictionaries, we run the danger of getting lost within the data. We will tidy up, and make safer, the code we will be writing, by creating a couple of `typealias`, which will also get Swift's type-checking magic working for us, ensuring that the data objects we pass around are really the objects we think they are.

Create a new Swift file, and call it `SharedConstants.swift`. Again, select both the phone and the WatchKit Extension targets. Now add the following code below the `import` statement:

```swift
typealias jsonDict  = [String: AnyObject]
typealias jsonArray = [AnyObject]
```

All we are doing here is adding two `typealias` declarations; `jsonArray` is an array of `AnyObject`, since we have no way of knowing in advance what exactly it will contain, and the `jsonDict` is a `Dictionary` of `String` keys and `AnyObject` values, which can also contain values of whatever type JSON throws at us.

Remember, although JSON has a limited number of types, it is still easy to make mistakes when handling JSON data—Swift considers a dictionary of `[String: String]` key-value pairs an entirely different creature to a dictionary of, say, `[String: [String]]` key-value pairs. Using `typealias` gets type-checking working in our favor.

## Introducing NSURLSession

Okay, so we have had a brief look at the data that will arrive onto the watch (which we will look at in more detail very soon), but how do we actually GET the data? How does the watch even connect to the Internet?

The answer to the latter question is, fortunately: very easily. Without us having to write a single line of code, the Apple Watch will happily decide whether it makes sense to use the iPhone's network connection, using what Apple calls Tetherless WiFi, or whether to connect directly to known networks to which the iPhone has connected in the past.

Be glad you don't have to write that code. Be very glad.

Apple's NSURLSession provides us with a huge amount of functionality (most of which we don't need here) around downloading and uploading data to and from the web, as well as caching that data.

# Disabling App Transport Security

In iOS 9, Apple introduced **App Transport Security** (**ATS**), which helps and encourages developers to use secure Internet Protocols whenever possible. Unfortunately, at the time of writing, http://openweathermap.org/ does not offer secure https URL's, so we'll need to get around ATS, otherwise the web requests will be refused by the operating system.

To do this, perform the following steps:

1. In the project navigator, locate the Info.plist file in the **Weather Watch WatchKit Extension**. Careful to select the right one (each target has one).

2. Right-click on it and from the contextual menu that opens, select **Open As | Source Code**. This is a much easier way to enter code than the default **Property List** view.

3. Add the following highlighted code under the first <dict> tag:

```
<?xml version="1.0" encoding="UTF-8"?>
<!DOCTYPE plist PUBLIC "-//Apple//DTD PLIST 1.0//EN""http://www.
apple.com/DTDs/PropertyList-1.0.dtd">
<plist version="1.0">
<dict>
    <key>NSAppTransportSecurity</key>
    <dict>
        <key>NSAllowsArbitraryLoads</key>
        <true/>
    </dict>
    <key>CFBundleDevelopmentRegion</key>
    <string>en</string>
```

Although it is possible to disable ATS on a per-domain basis, our app only uses the one server, and for the sake of simplicity we simply allow all non-secure http traffic, by adding this entry to the watch app target's property list.

# Using NSURLSession

So let's write some code that will download the data, using whatever connection the watchOS deems appropriate:

Create a new Swift file, call it `SessionManager`, and tick both the watch app and the phone app targets, then add the following code below the `import` statement:

```
let baseUrl = "http://api.openweathermap.org/data/2.5/
group?units=metric" //1
let APIKey = "78c75588dfe58276a694af9c660edxxx" //2
let faveCity = 2643741 //3
let cities = [faveCity, 2193733, 1273294, 5128581, 2950159, 3435910]
//4
```

These are the strings we will need to construct the URL of the data request. The comments in the code are as follows:

1.  This is the base URL with which all requests to `http://openweathermap.org/` begin. We will add more information to this depending on which data we want to request.

2.  Put your API key (which you got after creating the `http://openweathermap.org/` account) here, not forgetting the quote marks to turn it into a Swift `String`.

3.  Here we specify a favorite city — you can use this one (which is the City of London) or choose another from the list linked to above. This is the city whose data will be shown on the Glance screen.

4.  We create an array of city codes, using our `faveCity` code as the first.

Now we'll create the actual `SessionManager` class that will be available from elsewhere in the app. Add the following code to `SessionManager.swift`:

```
class WeatherSessionManager {

    static let sharedInstance = WeatherSessionManager() //5
    private init() {}

    var lastRequestDate: NSDate? //6
}
```

The comments in the code are as follows:

1. As we have done previously, we provide a static sharedInstance property so that the same instance is used throughout the app, and prevent the creation of a separate instance by making the init method private.

2. The lastRequestDate property will store the most recent date (and time) of a data request to the server. Any new request made will compare its own date with this one to establish whether or not to use the data already saved (or *cached*) by NSURLSession, or whether to fetch fresh data.

We'll now write the method that will create the full request URL string from the baseURL, APIKey, and cities strings. Add the following method to the SessionManager class:

```
func urlForCities(cityCodes: [Int]) -> String {
    var urlStr = baseUrl + "&APPID=" + APIKey + "&id=" //7
    for cityCode in cityCodes { //8
        urlStr += "\(cityCode),"
    }
    return urlStr //9
}
```

You will likely have many methods like this one in a larger, more complex app, so it is worth making doubly sure you understand what is going on. The comments in the code are as follows:

1. We concatenate the necessary strings using Swift String convenient + operator (if you're coming from Objective C, you will appreciate just how convenient it is). Note the request data is always preceded by "&", the name of the property (or field name) and a "=" character

2. We enumerate through the array of city codes, adding each to the urlString variable, and adding a comma (the comma after the last entry in the array is simply ignored).

3. Now that the full URL string is constructed, we return it to the calling function.

Now we get to the magic stuff. We will create and configure NSURLSession, and then call its session.dataTaskWithURL(url, completionHandler) method. This takes two arguments. The first is a straightforward NSURL, and the second is a function in the form of a closure.

Passing a function as an argument to another function may be new to you, and if it is, you're going to love it, I promise you. Being able to pass a function directly to another function means not having to maintain a direct relationship (such as a `delegate`) between two objects or not needing to bother with setting up notifications between them. Passing functions means, in effect:

```
doSomething(withSomeArg: ArgType, whenFinishedCallThisFunction:
FunctionType)
```

`dataTaskWithURL` stipulates that the function it is to call, once it is finished, must have the type (NSData?, NSURLResponse?, NSError?) -> Void, and so we'll create a `typealias` for this first, which will make the code look less intimidating. Add the following declaration to `SessionManager`:

```
typealias DataTaskCompletionHandler =
    (NSData?, NSURLResponse?, NSError?) -> Void
```

Now we can write a `fetchWeatherData` method. This will in turn call `session.dataTaskWithURL` method from `NSURL` class, but where do we get the `DataTaskCompletionHandler`?

That will be passed by the calling method to `fetchWeatherData` as its single argument.

Add the following method to the SessionManager class:

```
    func fetchWeatherData(completionHandler:
DataTaskCompletionHandler) {

}
```

If we had not created the `type alias`, this method signature would have looked like this:

```
    func fetchWeatherData(completionHandler: (NSData?, NSURLResponse?,
NSError?) -> Void) {

}
```

This is a little harder to read, though it is a pattern you'll quickly get used to.

Now add the following code to the `fetchWeatherData` method:

```
    func fetchWeatherData(completionHandler:
DataTaskCompletionHandler) {
        let sessionConfig =
NSURLSessionConfiguration.defaultSessionConfiguration() //10
```

```
          if let date = lastRequestDate
              where NSDate().timeIntervalSinceDate(date)   < 10.0 {
              sessionConfig.requestCachePolicy =
.ReturnCacheDataElseLoad
          } else {
              sessionConfig.requestCachePolicy = .UseProtocolCachePolicy
          }
          lastRequestDate = NSDate() //11
          let session = NSURLSession(configuration: sessionConfig) //12
          let apiCall = urlForCities(cities) //13
          if let url = NSURL(string: apiCall) {
              let task = session.dataTaskWithURL(url, completionHandler:
completionHandler) //14
              task.resume() //15
          }
      }
```

The comments in the code are as follows:

1.  Before we create NSURLSession, we need to prepare
    NSURLSessionConfiguration, and set its RequestCachePolicy according
    to how long it has been since the last request. The ten second value we have
    used here is to make testing it more convenient, but it will need changing to
    something more realistic later!

2.  We then update the lastRequestDate property with the date of this request,
    using a call to NSDate(), which returns the current date.

3.  We can now initiate the NSURLSession with the sessionConfig that we
    just created.

4.  Finally, we use the urlForCities method to construct the API call, and pass
    that as the argument to the NSURL init method, which creates the NSURL
    object that we need to pass to dataTaskWithURL.

5.  We also pass on the DataTaskCompletionHandler function that was passed
    in the call to fetchWeatherData.

6.  Nothing happens until we call resume method from NSURL class (there is no
    start method).

# Interface Controllers

Now the SessionManager class is ready to test. To do this, we simply need to
create the sharedInstance in the InterfaceController class which we will
now implement.

In the project navigator, select `InterfaceController.swift`, and add the following code to the `InterfaceController` class's `awakeWithContext` method:

```
override func awakeWithContext(context: AnyObject?) {
    super.awakeWithContext(context)
    requestWeatherData()
}
```

The compiler will complain about not knowing any such method, so we'll fix that now.

Add the following function to the `InterfaceController` class:

```
func requestWeatherData() { //1
    WeatherSessionManager.sharedInstance.fetchWeatherData() { //2
        (data: NSData?, response: NSURLResponse?, taskError:
NSError?) -> Void in
        if taskError == nil { //3
            do {
                let jsonData = try NSJSONSerialization.
JSONObjectWithData(
                    data!,
                    options:
                    .AllowFragments) as! jsonDict
                print(jsonData) //4
            }
            catch let jsonErrorasNSError {
                print(jsonError.localizedDescription)
            }
        }
    }
}
```

The comments in the code are as follows:

1.  We call the `fetchWeatherData` method of the `SessionManager` class method. Because we are passing a closure as the argument, we may place all of it, including curly braces, after the closing brackets between which arguments are usually placed. Without the details of the closure, the code looks like:

    ```
    WeatherSessionManager.sharedInstance.fetchWeatherData(){ ... }
    ```

2.  This closure will be called by the `NSURLSession` class's `dataTaskWithURL` method (inside the `fetchWeatherData` method) once it is finished, returning either the `data` and `response`, or the `taskError` as non-nil optional values.

3. If there is no error, we try to use the NSJSONSerialization class's JSONObjectWithData method to convert the NSData to a Swift Dictionary object that we can manipulate according to our needs. We wrap this in Swift's do-catch pattern: If something goes wrong, say, the data is not correctly formatted, then the catch block will be executed.

4. Assuming there are no problems, we print the jsonData object to the console.

# Testing in the console

To test the console, we must perform the following steps:

1. Make sure that your device or the watchOS Simulator is connected to a network (we will get no warnings if not, and the app will seem to have done nothing).

2. Select the **Weather Watch WatchKit App** scheme.

3. Hit **Run**.

The app will launch, and after a brief pause, the jsonData object will be printed to the console, where you can see the wealth of data that it contains. Later on, you may wish to take a longer look at the data and possibly add more of its contents to the app.

# Completing InterfaceController

We now need to complete the InterfaceController class. We will write the code first, and then prepare the UI in Interface Builder.

## Coding InterfaceController

You can delete the requestWeatherData method call from the awakeWithContext method, remember it was used in this case only for testing purposes and we don't need it any more.

Add the following code to the class:

```
class InterfaceController: WKInterfaceController {

    @IBOutlet var getWeatherButton: WKInterfaceButton!

    @IBAction func getWeatherButtonTapped() {
        requestWeatherData()
    }

    . . .
```

Here we are simply preparing the connections to the interface that we will create with Interface Builder.

Next add the following function to the class:

```
func showWeather(data: jsonDict) { //1
    pushControllerWithName("WeatherTableInterfaceController",
context: data)
}


func updateUIForNetworkActivity(isActive: Bool) { //2
    if isActive {
        getWeatherButton.setTitle("Fetching Data")
        getWeatherButton.setEnabled(false)
    } else {
        getWeatherButton.setTitle("Show Me the Weather")
        getWeatherButton.setEnabled(true)
    }
}
```

1. The `showWeather` method pushes a `WeatherTableInterfaceController` onto the screen.

2. Here we either disable the UI's only button while data is being fetched, and change the title to inform the user what is happening, or we re-enable it and change its title back to the original.

3. Finally, we must add a few more lines of code to our `requestWeatherData` method:

```
func requestWeatherData() {
    updateUI ForNetworkActivity(true) //3
    WeatherSessionManager.sharedInstance.fetchWeatherData(){
(data: NSData?, response: NSURLResponse?, taskError: NSError?) in
//4
        self.updateUIForNetworkActivity(false) //5
        if taskError == nil {
            do {
                let data = try NSJSONSerialization.
JSONObjectWithData(data!, options: .AllowFragments) as! jsonDict
                print(data)
                self.showWeather(data) //6
            }
            catch let jsonError as NSError {
                print(jsonError.localizedDescription)
            }
        }
```

```
                    else {  //7
                        let action = WKAlertAction(title: "OK", style:
    .Default, handler: {})
                        let alertText = taskError!.localizedDescription
                        self.presentAlertControllerWithTitle(
                            alertText,
                            message: "",
                            preferredStyle: .ActionSheet,
                            actions: [action])
                    }
                }
            }
```

4. We let the user see that the request has started

5. We make the call to the WeatherSessionManager class's fetchWeatherData method, passing in the closure that will be called by fetchWeatherData when it is finished.

6. Once fetchWeatherData is finished, we either get back an error or some data. Either way, we need to return the UI to its previous state of readiness.

7. If we get no error, then we try to convert the data to a jsonDict object. If that succeeds we call our showWeather method passing it the jsonDict.

8. If we get a non-nil taskError back, we inform the user of what went wrong with an alert sheet, presenting the NSError class's localizedDescription property to populate the alert with localized human readable text.

## Building the Interface

As promised, this is a really basic interface. You just need to follow the given steps:

1. In the project navigator, select **Weather Watch WatchKit App | Interface. storyboard**.

2. Drag a Button object onto the interface, and set both of its **Alignment** properties to **Center**.

3. Set the button's title to something appropriate, like "Show Me The Weather." Feel free to let your imagination run wild here.

4. Make sure the assistant editor is open (*Command-Option-Return*), and *Control-drag* from the button to @IBOutlet var getWeatherButton: WKInterfaceButton! in the source code.

5. Control-drag from the button to the @IBAction func getWeatherButtonTapped() in the source code.

And that's it, the UI is done.

# WeatherTableInterfaceController

Once the data is obtained and formatted as a `Dictionary`, it is passed to a new instance of `WeatherTableInterfaceController`. This class will then parse the data for whichever information it needs, and display it in a `WKInterfaceTable`.

## The simplicity of WatchKit tables

Using tables in watchOS is a lot simpler than in iOS, and although some of the concepts involved may be familiar, the fact is that `WKInterfaceTable` offers a much more restricted feature set, and is accordingly much simpler to handle.

Each row of the table is an instance of an `NSObject` subclass, which we will create. Any formatting of the row is done in the row object's `init` code (although in this app, we will not need to do anything here).

To populate the table with data, we simply override the `WKInterfaceController` class's `table(WKInterfaceTable, didSelectRowAtIndex: Int)` method.

There are no sections as we have in iOS's `UITableView`, there is no re-ordering of rows, there are no data source declarations, there is no need to declare a delegate. The considerable flexibility and power of iOS table views would not make a lot of sense on the small watch screen, and so we are dealing here with a much leaner creature.

It's all very simple. If you have implemented a `UITableView` before, you'll find this child's play.

## Coding WeatherTableInterfaceController

Let's start with the table row class, as follows:

1.  Create a new file, selecting the **watchOS | Source | WatchKit Class template**, and hit **Next**.

2.  Name it **WeatherTableRow**, make it a subclass of **NSObject**, and be sure to select **Swift** as the language. The settings should look like this:

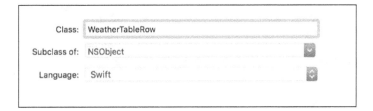

3. Save the file, making sure that just the Weather Watch WatchKit Extension target is selected (there is no need to leave the phone app target selected, it won't be using this class).

4. Add the following code to the `WeatherTableRow` class:

```
import WatchKit

class WeatherTableRow: NSObject {
    @IBOutlet var upperLabel: WKInterfaceLabel!
    @IBOutlet var lowerLabel: WKInterfaceLabel!
}
```

In the case of our table row, we do no more than add outlets for connection to the interface that we create in Interface Builder.

Now to the `InterfaceController` class itself:

1. Create a new file, selecting the **watchOS | Source | WatchKit Class template**, and hit Next.

2. Name it **WeatherTableInterfaceController**, make it a subclass of **WKInterfaceController**, and select Swift as the language. The settings should look like this:

We will start by adding the code necessary to populate the table with data provided in the form of our `WeatherData` typealias, which, you will recall, is an array of `WeatherSummary` objects.

Add the following code to the `WeatherTableInterfaceController` class:

```
class WeatherTableInterfaceController: WKInterfaceController {

    @IBOutlet var weatherTable: WKInterfaceTable! //1
    var weatherDataArray: jsonArray = [] //2
```

```
    override func awakeWithContext(context: AnyObject?) {
        super.awakeWithContext(context)
    }


    func loadTableData(data: WeatherData) { //3
        weatherTable.setNumberOfRows(data.count, withRowType:
"TableRowControllerID") //4
        for (index, summary) in data.enumerate() { //5
            if let row = weatherTable.rowControllerAtIndex(index) as?
WeatherTableRow {
                row.upperLabel.setText(summary.cityName)
                row.lowerLabel.setText(summary.summary)
            }
        }
    }


    override func table(table: WKInterfaceTable,
didSelectRowAtIndexrowIndex: Int) { //6
        self.pushControllerWithName("DetailsInterface", context:
weatherDataArray[rowIndex])
    }
}
```

The comments in the code are as follows:

1.  We will connect the **Table** object in Interface Builder to this `IBOutlet`.

2.  The `weatherDataArray` will store the data that is passed to the `WeatherTableInterfaceController` class's `context` argument.

3.  Once the data is ready for display (see below) we will call this method to load the table.

4.  We set the number of rows to the number of `WeatherSummary` objects contained in the `data` argument.

5.  We enumerate through the `data` array, and for each `WeatherSummary` object, we set the **Label** text properties of the row at the appropriate index.

6.  When the user selects one of the rows presenting the summary for a particular city, the table calls this method (which it will do with no code from us). We initiate an instance of the `DetailsInterface` (which we will code shortly), passing to it the data associated with that particular city.

# Parsing the JSON data

So let's now turn our attention to extracting the information we need from a Dictionary object (remember, we defined a typealias, jsonDict, to refer to objects of type [String: AnyObject]).

In order to do this, in this and in other projects that request JSON data from the internet, you will print the data to the console (as we have been doing already) in order to see how the data is structured - once you have decided which data you want, and where it is in the hierarchy of dictionaries, arrays, and objects, you can get coding.

Add the following code to the WeatherTableInterfaceController class:

```
func extractBasicWeatherData(data: jsonArray) -> WeatherData {
    var tableDataArray: WeatherData = [] //1
    for entry in data {
        if let //2
            name = entry["name"] as? String, //3
            weather = entry["weather"] as? jsonArray,//4
            summary = weather[0]["main"] as? String { //5
                let basicEntry = WeatherSummary(
                    cityName: name, summary: summary) //6
                tableDataArray += [basicEntry]
        }
    }
    return tableDataArray //7
}
```

The comments in the code are as follows:

1.  We initialize an empty WeatherData array, to which we can add WeatherSummary objects as we parse out their data.

2.  For each object in the data argument, we use what is called multiple optional binding; an if let statement and a comma-separated list of optional bindings that could potentially return nil:

3.  We check if a String with the key "name" exists.

4.  We check whether there is a jsonArray named "weather".

5.  We check whether the first object in that jsonArray, i.e. weather[0], contains a String object with the key "main".

6.  If all three conditions are met, meaning that none of these optional values are nil, we can construct a new WeatherSummary object and add it to tableDataArray we created above.

7. We return the array that contains the `WeatherSummary` objects we extracted from the `data` argument.

The last thing we need to do here is to add code to the `awakeWithContext` method, calling the methods we have implemented above.

Add the following code to the `awakeWithContext` method:

```
override func awakeWithContext(context: AnyObject?) {
    super.awakeWithContext(context)

    guard let data = context as? jsonDict else { print("Data is
not json dictionary"); return } //8
    guard let list = data["list"] as? jsonArray //9
else { print("No list data found"); return }

    weatherDataArray = list //10
    let basicWeatherData =   extractBasicWeatherData(list)
    loadTableData(basicWeatherData)
    }
```

1. Using Swift's `guard` keyword, we check first that the `context` argument is of type `jsonDict` (since the argument is specified to be of type `AnyObject`, it could be anything at all). If it is not, we log an error to the console and return from the function.

2. Similarly, we check whether that `jsonDict` contains a `jsonArray` with the key `"list"`. If not, we log the error and return from the function.

3. If the list object is not nil, we copy it to the `weatherDataArray` property, for passing to the `DetailsInterface` when the user selects a city from the table.

# Building the WeatherTable UI

The interface for the `WeatherTableInterfaceController` is nearly as simple as the one we built for `InterfaceController`:

1. Drag an `InterfaceController` object into the storyboard window, placing it to the right of the `InterfaceController` we have already finished.

2. In the Attributes Inspector (*Command-Option-4*), set the new `InterfaceController` Identifier to `WeatherTableInterfaceController`.

3. In the Identity Inspector (*Command-Option-3*), set the new `InterfaceController` Class to `WeatherTableInterfaceController`.

4. Now drag a Table object onto the `WeatherTable` interface. With the Assistant Editor visible, *Control-drag* to `@IBOutlet var weatherTable: WKInterfaceTable!` in the source code, as illustrated here:

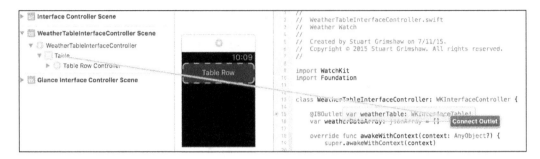

5. Set the Table Row Controller Class to **WeatherTableRow** in the Identity Inspector, as illustrated here:

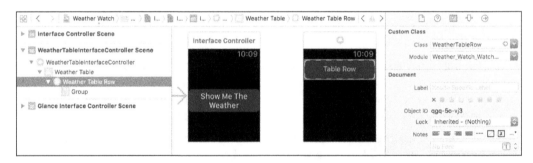

6. In the Attributes Inspector, set **Weather Table Row's Identifier** to **TableRowControllerID**.

7. Select the Group that is contained in **Weather Table Row**.

8. Set its **Layout property** to **Vertical**, and its **Height property** to **Size To Fit Content**.

9. Drag two **Label** objects onto the group. With `WeatherTableRow.swift` visible in the **Assistant Editor**, Control-drag from the **Label** objects to the two `IBOutlet` in the code, as illustrated here:

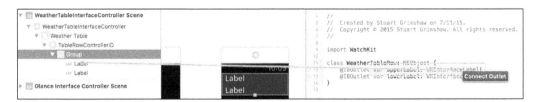

# Run the app

At this point, if you run the app, you should be able to load the data, and see it displayed in the `WeatherTable`. This is surely one of the most satisfying parts of developing a web-connected app, watching the world's data pour into the device and being presented for your delectation!

Selecting a city in the table won't get you anywhere yet, however, because we have not yet created the `DetailsInterfaceController` with ID `DetailsInterface` that the code calls. So we'll take care of that next.

# DetailsInterfaceController

Here, we want to display to the user a more detailed version of the weather data, and a look at the `Dictionary` that we created from the JSON data will reveal which data is available.

## Coding the DetailsInterfaceController

Most of the implementation here is familiar territory, though we will pause to have a closer look at what it takes (which is not much, you'll be pleased to know) to get an image from the web and display it along with the text.

1.  Create a new file, selecting the **watchOS** | **Source** | **WatchKit** Class template, and hit **Next**.

2.  Name it `DetailsInterfaceController`, make it a subclass of `WKInterfaceController`, and select Swift as the language.

3.  Add the following code to the `DetailsInterfaceController` class:

    ```swift
    @IBOutlet var detailsLabel: WKInterfaceLabel!
    @IBOutlet var image: WKInterfaceImage!

    let imageBaseUrl = "http://openweathermap.org/img/w/"

    override func awakeWithContext(context: AnyObject?) {
        super.awakeWithContext(context)
        if let data = context as? jsonDict {
            displayData(data)
        }
    }
    ```

4.  There is nothing here that you have not seen before. The compiler will complain that it doesn't know any `displayData` method, so we'll fix that next.

5. Add the following function to the `DetailsInterfaceController` class:

```swift
func displayData(data: jsonDict) {
    var detailsText = ""
    if let cityName = data["name"] as? String{
        detailsText += cityName + "\n"
    }

    if let main = data["main"] as? jsonDict {
        if let humidity = main["humidity"] as? Int { //1
            detailsText += "Humidity: \(humidity)\n"
        }
        if let temp = main["temp"] as? Int {
            detailsText += "Temp: \(temp)\n"
        }
    }

    if let
        wind = data["wind"] as? jsonDict,
        speed = wind["speed"] as? Double
    {
        detailsText += "Windspeed: \(speed) Km/h"
    }

    if let weather = data["weather"] as? jsonArray {
        if let icon = weather[0]["icon"] as? String {
            fetchIcon(icon)
        }
        if let descriptionStr = weather[0]["description"] as?
String {
            detailsText += "Weather: " + descriptionStr + "\n"
        }
    }

    detailsLabel.setText(detailsText)
}
```

Although this is all stuff we've seen previously, do take the time to read carefully through it—parsing JSON data is a critical part of iOS development (and by extension, therefore, watchOS development), and Swift's type-safe way of handling it needs to become something you can do in your sleep.

The comments in the code are as follows:

1. Note that we're handling a few `Int` and `Double` objects here—as we saw above, JSON isn't all strings and dictionaries.

2. We extract the name of the icon image that the server includes in the JSON data, and pass it to the `fetchIcon` method, which we will implement next.

3. Add the following function to the `DetailsInterfaceController` class:

```
func fetchIcon(iconStr: String) {
    let imageUrlStr = imageBaseUrl + iconStr + ".png"
    if let imageUrl = (NSURL(string: imageUrlStr))
    {
        let sessionConfig= NSURLSessionConfiguration.
defaultSessionConfiguration()
        sessionConfig.requestCachePolicy=
.ReturnCacheDataElseLoad
        let session = NSURLSession(configuration:
sessionConfig)
        let task = session.dataTaskWithURL(imageUrl){ //1
        (
            data: NSData?,
            response: NSURLResponse?,
            error: NSError?) in
            if let unwrappeData = data { //2
                self.image.setImage(UIImage( //3
                    data: unwrappeData))
            }
        }
        task.resume() //4
    }
}
```

The comments in the code are as follows:

1. Once again, we use the `dataTaskWithURL` method from the `NSURLSession` class to obtain the data we need, but this time instead of JSON data, we are getting image data returned to us.

2. We check that data was returned by the request to the server.

3. We pass this `NSData` to `UIImage init(data:)` convenience `initializer` method.

4. Remember, nothing happens until we call the `resume` method of `dataTask` class

# Creating the DetailsInterfaceController UI

For this screen we need only add an Image object and a Label object to the **DetailsInterface** Scene. To do this, perform the following steps:

1.  In the project navigator, select `Interface.storyboard`.

2.  Drag an **InterfaceController** object into the Interface Builder window, and in the Identity Inspector, set its **Custom Class** to `DetailsInterfaceController`.

3.  In the Attributes Inspector, set the Identifier to **DetailsInterface**. (Make sure you hit **Return** to enter the new name before navigating away from it—this one bites me every time!)

4.  Drag an Image object onto the `InterfaceController`, and connect it to @ `IBOutlet var image: WKInterfaceImage!` in the source code.

5.  Set its **Image Mode** property to **Aspect Fit**.

6.  Set its **Width** property to **Relative To Container**, with a value of `0.5`, and its **Height** to **Relative To Container** with a value of **0.2**.

7.  Set its **Horizontal Alignment** property to **Center**.

8.  Drag a Label object onto the **InterfaceController**, and connect it to @ `IBOutlet var detailsLabel: WKInterfaceLabel!` in the source code.

9.  Set its **number of Lines** property to `0`, so that it will display as many lines of text as necessary.

10. Set its **Horizontal Alignment** property to **Center**.

11. Set its **Width** property to **Relative To Container**.

# Run the complete app

The main app is now complete, and when you run it, you should have access to not only the weather summaries of a number of cities, but also the detailed weather data of the cities you select from the table.

You could, of course, try experimenting with the presentation of detailed information, adding perhaps more of the data that is available in the JSON returned from the server.

# Glances

We have the option of adding a Glance scene to a watch app, and while many will not need it, our app is going to make use of this opportunity to provide a small snippet of the information provided by the main app. The Apple Watch's Glance is reachable at all times with a swipe upwards from the watch face, and is intended as a brief, non-interactive engagement with your app. Given how easy it is to reach, the Glance enables your app to provide a limited amount of data which is immediately accessible.

You cannot add most UI objects to a Glance scene, there will be no tapping buttons and selecting from tables and the like. If your user wishes to engage more deeply with your app, she can do so by tapping the Glance screen which will open your main app, just as if she had launched it from the home screen.

When designing and implementing the Glance screen, the main challenge is not so much deciding how to code it, which is likely to be a simplified version of some part of the main app, but much more deciding exactly which information your users are likely to need in the context of a short look at your app. Each app will have different needs, of course, but the general approach is to identify the most important, core purpose of your app. Which is the feature that makes most sense without the rest of the app's functionality?

We will be presenting the information about the city that we identified as `faveCity` in our code in our Glance.

# Coding the GlanceController

There is little to do in the `GlanceController` that we have not done before, the code is basically a reduced version of the `WeatherTableInterfaceController` class. We only need to add the code and hook it up to a very minimal UI.

In the project navigator, select the file `GlanceController.swift`, and replace everything inside the `GlanceController` class's curly braces with the following code:

```
@IBOutlet var topLabel: WKInterfaceLabel!
@IBOutlet var bottomLabel: WKInterfaceLabel!

override func awakeWithContext(context: AnyObject?) {
    super.awakeWithContext(context)
    requestWeatherData()
}
```

```
func requestWeatherData() {
    WeatherSessionManager.sharedInstance.fetchWeatherData(){
(data: NSData?, response: NSURLResponse?, taskError: NSError?) in

        if taskError == nil {
            do {
                let jsonData = try NSJSONSerialization.
JSONObjectWithData(data!, options: .AllowFragments) as! jsonDict
                self.showWeather(jsonData)
            }
            catch let jsonErrorasNSError {
                print(jsonError.localizedDescription)
            }
        }
        else {
            let action = WKAlertAction(title: "OK", style:
.Default, handler: {})
            let alertText = taskError!.localizedDescription
            self.presentAlertControllerWithTitle(
            alertText,
            message: "",
            preferredStyle: .ActionSheet,
            actions: [action])
        }
    }
}

func showWeather(data: jsonDict) {
    guard let list = data["list"] as? jsonArray else {return }
    if let
        cityName = list[0]["name"] as? String,
        weather = list[0]["weather"] as? jsonArray,
        descriptionStr = weather[0]["main"] as? String {
            topLabel.setText(cityName)
            bottomLabel.setText(descriptionStr)
    }
}
```

# Building the GlanceController UI

Again, although there is nothing new here, do read carefully through the code, making sure that you understand exactly what is going on. This repetition of common tasks in programming is what will turn you into a better and faster developer (those two adjectives are in order of importance, by the way).

# Building the Glance interface

We will now turn to setting up the UI in Interface Builder.

 Don't worry if you didn't include the Glance Scene when you set up the project—you can simply drag a **Glance Interface Controller** object from the Object Library (*Command-Option-Control-3*).

The **Glance Interface Controller** currently looks like this:

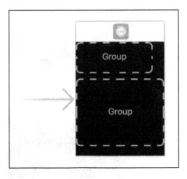

In the storyboard window, select the `GlanceController` that Xcode included when we first created the project, and perform the following steps:

1. With **the Glance Interface Controller** Scene selected in the **Document Outline**, the **Attributes Inspector** shows the two Groups provided by the template:

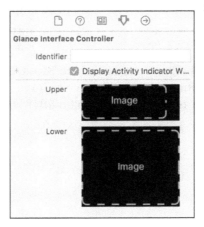

2.  We don't need Image objects here, we want **Label** objects, so click on Upper and select the Label that is shown as "**Label.............**" (you might have to scroll down to see it).

3.  Do the same for **Lower**, but select the **Label** and Footer combination, so that the Attributes Inspector looks as it does here:

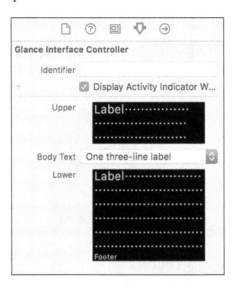

4.  Select the top **Label**, and set its **Title** to **Fetching Data**.

5.  Connect it to @IBOutlet var topLabel: WKInterfaceLabel! In the source code.

6.  Select the middle **Label** and set its **Title** to be empty.

7.  Connect it to @IBOutlet var bottomLabel: WKInterfaceLabel! In the source code.

8.  Select the **Footer Label**, and set its **Title** to **More Cities...** (this label, while not being anything interactive, will serve to invite the user to tap on the screen to get to your main app—it won't matter where he taps, though).

Your storyboard should now look like this:

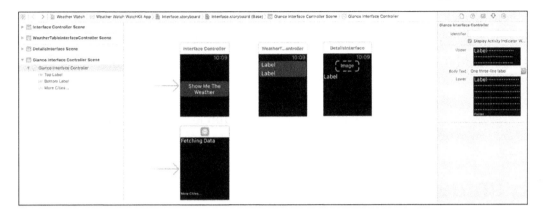

# Running and testing the Glance

To test the Glance Scene, select the **Glance Weather Watch WatchKit** App scene, as illustrated here:

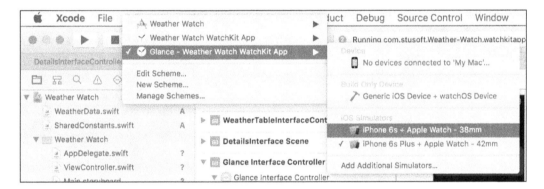

Now hit **Run**, and you'll see your Glance Scene launch on the Simulator or on your physical device, as if you had just swiped up from your watch face (assuming your app is selected to **Show In Glances** in the iOS Watch app).

# Challenges for expansion

If you are inclined to take further what you have learned in this and previous chapters, you may wish to expand and improve the app in the following ways:

- Add the same functionality to the iPhone app, but present much more of the wealth of information provided by http://openweathermap.org/.

- The UI is a very Spartan one, you should be able to find any number of tweaks that will make your version of the app look a thousand bucks more expensive.

# Summary

Wow, what a chapter. We have seen how to obtain open source data using NSURLSession, including composing URL's and caching the results to reduce data traffic, we have seen how to format the data returned by our requests and parse it to obtain the information we need, and then how to present that information in a table. We have used the console to inspect returned data, added user alerts when errors occur, and pushed WKInterfaceController programmatically, as a more flexible form of navigation. Finally, you have added a Glance Scene to your app.

In the next chapter, we will be using another (similarly advanced) cornerstone of WatchKit development: Core Location. These are mobile devices, after all!

# 7
# Plot Buddy – All about Location

Naturally, one of the most important features of mobile devices is their ability to find their own location, whether using GPS sensors or cell data tower locations, or whatever. The Apple Watch is quite capable of capturing location data, either using the iPhone's GPS or when the iPhone is not present, translating its accelerometer data into location data. A pretty neat trick, from a development perspective, and one which enables the user to leave the phone where it is and take off with just the watch.

In developing this app, we will look at several new aspects of programming with the WatchKit framework, as well as a couple of techniques that apply equally to watchOS and iOS development. They include the following topics:

- Property Lists
- Core Location framework
- Protocols
- Adding an init method
- Forcing named arguments

## Planning the app

Although we will be using some pretty advance programming for this app, the actual use scenario is quite straightforward and our user story will be simply structured.

# Mission Statement

**Plot Buddy** will enable the user to store any number of locations directly from the watch; while moving around, say, a plot of land, or a regular jogging route, or any number of situations in which a list of related location data would be useful. When the user is finished, it will be possible to send the data to the paired iPhone (once it is in range), from where it could be used in any number of ways.

# User Story

After launching the app, the user will be offered a single active button with the title **Start Plotting**, which will allow him to start plotting locations.

Once plotting has started, a second button, titled **Add Current Coordinates**, will become active which will trigger the addition of the current location's geo-coordinates to a list that is maintained by the app. The location coordinates, assuming the location is found, will be displayed on screen (or an error notification if no location data is retrieved). The Start Plotting button will now have the title **Stop Plotting**.

This will continue until he stops plotting at which point he will see a button, titled **Show Plots**. This will offer the chance to inspect all the plots he has stored in the current list presented in a table on a new screen.

Each item in the table will display the number of the location in the list as well its latitude and longitude values.

From the main screen, a Force Touch will display a menu that will enable the user to send all the current data to the iPhone app.

# Setting up the project

So the actual flow of our user experience is very simple and is linear in nature – the user collects data, stores, and sends it to the iPhone, there are no branches in the user's behavior.

# Requirements

We will need to make use of the following frameworks provided by WatchKit:

- Core Location
- Watch Connectivity

We have seen in previous chapters how to leverage Watch Connectivity and we will see shortly the few simple steps we need to take, in order to access core location.

As far as custom structures are concerned, we need to create the following classes:

- Shared constants, to facilitate data exchange between the watch and the phone
- A Watch Connectivity manager, as we have created in previous projects
- A location manager to fetch and store location data and make it available to the interface controllers
- The main user interface and its associated `WKInterfaceController` class
- An interface and its `WKInterfaceController`, to present the plotted data in table form
- The table row controller, to format the appearance of the displayed data

Begin by downloading the Plot Buddy Template folder from here (use the Download ZIP link):

`https://github.com/codingTheHole/BuildingAppleWatchProjectsBook`

# Data structure

Firstly, we define a `typealias` for the data structure added to both the watch and the phone targets, that will hold the location data, by following these steps:

## Shared constants

In the downloaded project, create a new Swift file, name it `SharedConstants.swift`, being sure to select both targets, **Plot Buddy** and **Plot Buddy WatchKit Extension**, as illustrated here:

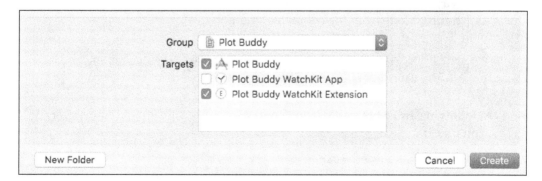

Select this file in the project navigator and replace the `import` statement with the following code:

```
Import WatchKit

typealias LocationSet = [CLLocation] //1

let ApplicationContextDataKey = "Data" //2
```

1. Our `LocationSet` data structure is nothing more than an array of `CLLocation` objects, which is the type returned by our calls to the Core Location framework.
2. For the sake of safety, we define a String constant for use as the `Key` to the data object passed from watch to phone in the `ApplicationContext`.

# Getting location data

Now we get to play with some new stuff.

Apple's Core Location framework provides a wealth of functionality around tracking a user's location. Although we will only scratch the surface in this app, you will see that a `CLLocation` object contains a set of data about the user's geo-coordinates and time-zone, speed of motion, and a time-stamp.

If we wish to access location data, we need to ask permission of the user before we can get a response to our requests to the operating system. And to do this we must first inform the system of our app's need to do so. We do this by adding an entry into the iPhone's `Info.plist` file.

## Modifying the iPhone's Info.plist

The easiest way to do this is to *Control-click* the file in the project navigator and select **Open As | Source Code** from the contextual menu.

 Be careful, don't confuse this with the watch app's identically named `Info.plist` file!

Add the following entry to the XML text contained in the file, immediately after `<dict>`, on line 4:

```
<?xml version="1.0" encoding="UTF-8"?>
<!DOCTYPE plist PUBLIC "-//Apple//DTD PLIST 1.0//EN""http://www.apple.
com/DTDs/PropertyList-1.0.dtd">
```

```
<plistversion="1.0">
<dict>

    <key>NSLocationWhenInUseUsageDescription</key>
    <string>Location Buddy requires location data for its core
functionality</string>

    <key>CFBundleDevelopmentRegion</key>
    <string>en</string>
...
```

You can replace the text between the `<string>` and `</string>` tags with whatever message you would like your user to see when watchOS asks for permission to use the location data. Messages like this should be tailored to your expected user group, you'll want to adopt an appropriate tone, *Hey, gonna need your location to do my job*, would be great for a young, hip target user group, for the corporate world not so much.

[  Bear in mind that the system will not ask the user again if he declines, so better keep it informative and persuasive. ]

With this entry in the dictionary in place, we are ready to code the class that will handle the requests to the operating system for the location data, the callbacks that the system will make to return that data, and the app-internal logic around making the data accessible.

# Creating PBLocationManager

As usual, we will encapsulate this functionality into a class of its own. There is no need to add this to the `WKInterfaceController`, as is so often done, since a view controller need know nothing about the implementation details around accessing the data.

## Create the Class

With the **Plot Buddy WatchKit Extension** selected in the project navigator, create a new Swift file and call it **PBLocationManager**. The PB prefix, and indeed the whole name, could be anything at all, but the app's name is Plot Buddy, so that kind of makes sense. The **Plot BuddyWatchKit Extension** should be automatically selected as the target, but better check anyway, and there is no need to select the **Plot Buddy** iPhone target selected.

Replace the `import` statement with the following code:

```
Import WatchKit

class PBLocationManager: NSObject, // 1
CLLocationManagerDelegate //2
{

}
```

The code is explained below:

1. We need to make the class a subclass of `NSObject` in order to be able to make it conform to the `CLLocationManagerDelegate` protocol.
2. We need to conform to the `CLLocationManagerDelegate` protocol in order to receive data from lore location.

# Delegates and Protocols

Although an in-depth discussion of protocols and the delegate pattern is beyond the scope of this book, it is worth a brief pause to take a high level look at what these techniques mean for the design of our app.

Basically, we are looking for a way for one class to communicate with another. This simply means that when `Class A` sends a method call to `Class B`, it must be sure that `Class B` is expecting to get such a call and offers some sort of implementation. `Class A` doesn't know or care how `Class B` does it; it just needs to know that the method will exist. Indeed, Swift's strong typing mechanism will prevent your code from even compiling if, for example, `Class A` calsl a method `doSomethingClever()` on `Class B` and `Class B` has no such method.

By declaring `Class B` to be conformant to a particular protocol, we are promising to implement a set of methods that is defined by a protocol definition. And now we can tell `Class A` that `Class B` conforms to that protocol, so that `Class A` (and Xcode itself) can be sure that it will not be calling non-existent methods.

So a protocol definition will look something like this:

```
protocol someProtocol {
    func doSomethingClever()
}
```

And a class that conforms to the protocol will look similar to the following:

```
Class ClassB: someProtocol{
    func doSomethingClever() {
        . . .
    }
}
```

So, as far as our app goes, we can tell the PBLocationManager class (when we get to it) that there will be an object (to which we will store a reference called delegate), which will implement the methods contained in a protocol (which we will call PBLocationManagerDelegate), by which PBLocationManager can pass on any data it needs to.

# Define the protocol

Although we could put the definition of the PBLocationManagerDelegate protocol anywhere, we will add it to the PBLocationManager.swift file.

Add the following code to the file below the import statement:

```
Protocol PBLocationManagerDelegate {
    func handleNewLocation(newLocation: CLLocation)
    func handleLocationFailure(error: NSError)
}
```

Now, any class can adopt the protocol and will be forced to implement the two methods declared in the definition above. Our interface controller will do just that.

# Implement PBLocationManager

Now we can code the PBLocationManager class itself. Start by adding the following properties to the class:

```
Class PBLocationManager: NSObject, CLLocationManagerDelegate
{
    Let locationManager = CLLocationManager() //1
    var currentLocations = LocationSet() //2
    var delegate: PBLocationManagerDelegate //3
}
```

The comments in the code are as follows:

1. We create an object that is an instance of CLLocationManager, part of core location, which will do the actual communication with the operating system.

2. We create an empty array of type LocationSet, which we defined in SharedConstants.swift. This will hold the data as it is returned by our locationManager object.

 Note that CLLocationManager and PBLocationManager are two entirely unrelated classes. One is supplied by WatchKit, one we have defined ourselves.

3. We declare a property that will be some object that conforms to the PBLocationManagerDelegate protocol. Since we know that it implements the necessary methods, we don't care what type it is (we'll just make function calls and supply arguments, to be dealt with as the delegate sees fit).

The compiler now complains that the PBLocationManager class has no initializers. So what's this about?

Well, up until now, when we have declared the properties of a class, we have also given each property a value. This is not the case with the delegate property, which will be some object outside of the class that doesn't exist at compile time, so we can't assign it a to our property. But Swift does not permit an object, including our PBLocationManager class, to be initiated with unassigned properties (unless they are optionals, but this class is useless without a delegate to talk to, so we insist on it having one by forcing its assignment during initialization).

So we must write in init method that will make sure a delegate object is provided. Add the following method to the class:

```
init(delegate: PBLocationManagerDelegate) { //1
    self.delegate = delegate //2
    super.init() //3

    locationManager.delegate = self //4
    locationManager.desiredAccuracy = kCLLocationAccuracyBest //5
    if CLLocationManager.authorizationStatus()
== .NotDetermined{ //6
        locationManager.requestWhenInUseAuthorization()
    }
}
```

The comments in the code are as follows:

1.  The `init` method for this class requires an object that conforms to the `PBLocationManagerDelegate` protocol.

2.  The object passed with the call to `init` is assigned to the `delegate` property of the class.

3.  When all properties of the class have some value assigned to them (and only then) we can call the `init` method on the class's `superclass` (in this case, `NSObject`).

4.  Now we can begin to configure the `locationManager` object, setting its own `delegate` property (nothing to do with our class's) to be `self`, so that we get the location data returned to us in the methods defined by that protocol, the methods of which our class will need to implement.

5.  Core Location can provide location data with varying degrees of accuracy: Highest accuracy — most data — most processing power and bandwidth, but for this app, this is the appropriate setting. Not something you'd need for, say, a street-finder app.

6.  If the app has not yet asked the user for permission to access the location data, it will do so now. This means that the alert pops up on app launch, before the user begins to collect data.

Next, we need to write the code that actually requests the location data from the system. We include a check of the `CLLocationManager.authorizationStatus` to make sure that the user has indeed authorized access to the data and, if not, we log this to the console (this is for debugging purposes, since the user will not see the console output). Add the following method to the `PBLocationManager` class:

```
func requestLocation()
{
    let authorizationStatus = CLLocationManager.
authorizationStatus()
    switch authorizationStatus {
    case .NotDetermined:
        print("requestLocation = NotDetermined")
    case .AuthorizedWhenInUse:
        locationManager.requestLocation()
    case .Denied:
        print("requestLocation = Denied")
    default:
        print("requestLocation = default")
    }
}
```

So much for sending the request, but how do we get the answer? Protocols and delegates are once again the answer. We have declared PBLocationManager to conform to the CLLocationManagerDelegate protocol, and we have set our class to be such a delegate, so now we can simply implement the necessary methods, which will be called by the CLLocationManager once its work is done.

 CLLocationManagerDelegate protocol lists a number of methods, which are all declared to be optional since classes will have different requirements of Core Location. This is why the compiler has not complained that we have not implemented any such methods. Of the methods declared, we are interested in just those pertaining to the success or failure of obtaining location data.

Add the following two methods to the PBLocationManager class:

```
func locationManager(
    manager: CLLocationManager,
    didUpdateLocations locations: LocationSet) {
    currentLocations += [locations[0]] //1
    delegate.handleNewLocation(locations[0]) //2

}
func locationManager(
    manager: CLLocationManager,
    didFailWithError error: NSError) {
    delegate.handleLocationFailure(error) //3
}    }
```

The comments in the code are as follows:

1. If Core Location calls this delegate method, then we are interested only in the first location of the array (there will only be one location each time the method is called anyway), which we add to the class's currentLocations array.

2. The delegate will be the WKInterfaceController subclass that we will create later. It will need to know what the new location is, in order to display its coordinates on screen.

3. Similarly, if no location data is available, the user interface will need to be updated accordingly and so we pass the error onto the delegate WKInterfaceController subclass, which controls our UI.

Finally, we add a method that other classes can call to reset the `currentLocations` array to an empty state. Add the following to the class:

```
func clearLocations() {
currentLocations = []
    }
```

And that's it for our `PBLocationManager` class. It would be a splendid idea to test the code before we continue with development and this we can do with a couple of lines of code once we have created the `InterfaceController` that will control our main screen

# The Interface Controllers

With the `PBLocationManager` in place, we have done the heavy lifting of the app. The interface controllers are very slim in design and simple in implementation.

## Create the InterfaceController class

Select the `InterfaceController.swift` file that was created as part of the Xcode template and replace all of the code in it (including the `import` statement) with the following:

```
Import WatchKit

class InterfaceController: WKInterfaceController,
PBLocationManagerDelegate {

    var locationManager: PBLocationManager! //1

    override func awakeWithContext(context: AnyObject?) {
        super.awakeWithContext(context)
        locationManager = PBLocationManager(delegate: self) //2
        locationManager.requestLocation() //3
    }

    func handleNewLocation(newLocation: CLLocation) { //3
        print(newLocation)
    }

    func handleLocationFailure(error: NSError) { //4
        print(error)
    }

}
```

What we are doing here is writing the bare minimum code necessary to test the PBLocationManager class. Some of this code we will remove once we are satisfied that the Core Location data is successfully being obtained.

The comments in the code are as follows:

1. We declare a property, locationManager, that is an instance of our PBLocationManager class. By using the exclamation mark, we are creating an implicitly unwrapped optional, which basically means that instead of creating an optional value and testing that it is not an empty optional when we need to use it, we promise the compiler that it will exist by the time we call it.

2. How can we promise this? We initialize an instance in the awakeWithContext method, by which we can be sure that locationManager never equals nil. If it did, the app would crash.

3. We fire off a call to the locationManager to request the current location.

4. If the locationManager receives its data, it will call this method on its delegate. For the moment we just print the results to the console for debugging

Have a good look through this little snippet of code; it's something you'll be doing quite often as a developer. It is much better to ensure that the code is working as you expect it to now, in isolation, rather than later when there is a ton of other code that could also be responsible for things not working out as they should.

# Test in the console

Okay, we're nearly ready to run a test, but first we need to select the right Scheme to the right of the **Run** button in Xcode; we need the **Plot Buddy WatchKit Extension** Scheme.

Now hit **Run**.

> If nothing happens, check that you have entered the NSLocationWhenInUseUsageDescription XML entry into the Info.plist of the iPhone app, not the WatchKit app.

Your Watch Simulator or your device should now ask you for permission to access your location data, as pictured here:

Whether or not you press **Dismiss** doesn't really matter, what counts is that you allow access on your iPhone or Simulator, as illustrated below:

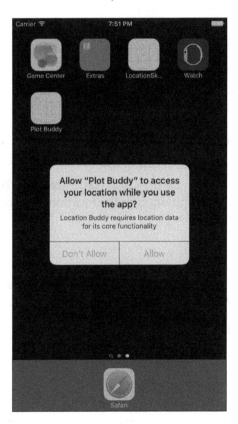

It will come as no surprise to you that you should select **Allow**.

When you do so, the dialog box will also disappear from the watch screen. But nothing more will happen yet—the call to requestLocation came and went long before you tapped on **Allow**.

So let's run it again. This time, the request for data is made. After a few seconds, the response that you've logged to the console (nothing will appear on the watch screen yet) may or may not look something like this:

```
<+37.33233141,-122.03121860> +/- 5.00m (speed 0.00 mps / course -1.00)
@ 11/23/15, 8:16:54 PM New Zealand Daylight Time
```

If it does, congratulations! You have successfully harvested a bunch of location data from watchOS. If the response you get is this:

```
Error Domain=kCLErrorDomain Code=0 "(null)"
```

then you are probably using the Watch Simulator and you'll need to let it know where you would like the Simulator to pretend to be: Go to Watch Simulator, select **Debug | Location** and choose **Apple** from the list of location data that the simulator can fake for you. Run the app again.

## Beware of the glitches

At the time of writing, Xcode is still displaying some growing pains when it comes to the Watch Simulator. It may take several attempts to get it to accept the new location you select, so you may have to alternate between **Apple** and **None** a couple of times before the dreaded null response stops plaguing you.

When testing the location code you may also come across cached responses, which will show an incorrect time stamp, though the location is correct. Don't let this bother you, on the device it all works fine.

## Code

Okay, now that we have PBLocationManager doing what it should, we can dispense with the test code and start to implement InterfaceController for real.

We start off with some string constants that will specify button titles and label text. Add the following code after the import statement, but before the InterfaceController class definition:

```
Let kStartPlotting = "Start\nPlotting"
Let kStopPlotting = "Stop\nPlotting"
Let kStoreCoordinates = "Add Current\nCoordinates"
Let kFetching = "Fetching Location..."
Let kLocationFailed = "Location Failed"
```

The implementation of the class contains very little that you have not seen before. Replace the entire `InterfaceController` class code with the following:

```
Class InterfaceController: WKInterfaceController,
PBLocationManagerDelegate {

    var locationManager: PBLocationManager!
    var isRecording = false

    @IBOutlet var startStopGroup: WKInterfaceGroup!
    @IBOutlet var addPlotGroup: WKInterfaceGroup!
    @IBOutlet var showPlotsGroup: WKInterfaceGroup!
    @IBOutlet var showPlotsButton: WKInterfaceButton!
    @IBOutlet var infoLable: WKInterfaceLabel!
    @IBOutlet var startStopButton: WKInterfaceButton!
    @IBOutlet var addPlotButton: WKInterfaceButton!

    Override func awakeWithContext(context: AnyObject?) {
        super.awakeWithContext(context)
        locationManager = PBLocationManager(delegate: self)
        updateUI(running: false)
    }

    @IBAction func startStopButtonPressed() {
        toggleRecording()
    }
    @IBAction func addPlotButtonPressed() {
        fetchLocation()
    }

    @IBAction func showPlotsButtonTapped() {
        pushControllerWithName("PlotsScene", context: locationManager.
currentLocations)
    }

    @IBAction func sendPlotsButtonTapped() {
WatchConnectivityManager.sharedManager.
sendDataToWatch(locationManager.currentLocations)
    }

    func toggleRecording() {
        isRecording = !isRecording
        if isRecording {
            locationManager.clearLocations()
```

```
            }
        updateUI(running: isRecording)
    }


    func fetchLocation() {
        updateUI(fetching: true)
        locationManager.requestLocation()
    }


    func handleNewLocation(newLocation: CLLocation) {
        addPlotButton.setTitle(kStoreCoordinates)
        infoLable.setText("Lat:\n" + "\(newLocation.coordinate.
latitude)" + "\nLong:\n" + "\(newLocation.coordinate.longitude)")
updateUI(fetching: false)
    }


    func handleLocationFailure(error: NSError) {
        infoLable.setText(kLocationFailed)
        updateUI(fetching: false)
    }


    func updateUI(running running: Bool) {

        infoLable.setText("")
        infoLable.setHidden(!running)
        startStopButton.setTitle(isRecording ? kStopPlotting :
kStartPlotting)

        showPlotsGroup.setHidden(running)
        showPlotsButton.setEnabled(locationManager.currentLocations.
count>0)
showPlotsGroup.setBackgroundColor(locationManager.currentLocations.
count>0 ? UIColor.blueColor() : UIColor.clearColor())

        addPlotGroup.setHidden(!isRecording)
    }


    func updateUI(fetching fetching: Bool) {
        infoLable.setHidden(fetching)
```

```
        startStopButton.setEnabled(!fetching)
        startStopGroup.setBackgroundColor(fetching ? UIColor.
    clearColor() : UIColor.blueColor())

        addPlotButton.setEnabled(!fetching)
        addPlotGroup.setBackgroundColor(fetching ? UIColor.
    clearColor() : UIColor.blueColor())
        addPlotButton.setTitle(fetching ? kFetching :
    kStoreCoordinates)
    }
}
```

Don't worry about the compiler error, this is happening because we have not yet added the `WatchConnectivityManager` class.

This class implements the methods we specified in our `PBLocationManagerDelegate` protocol definition.

```
func handleNewLocation(newLocation: CLLocation)
func handleLocationFailure(error: NSError)
```

This is the promise we make when we declare this class to conform to the protocol.

It's worth pointing out the method signatures of the last two methods. Why is the argument name written twice?

```
func updateUI(fetching fetching: Bool)
```

This forces the code that calls the method to explicitly name the argument being passed in this form:

```
updateUI(fetching: false)
```

This is much clearer, and makes understanding the code (when you return to it a year from now) very much easier, and is more elegant than including the name of the argument in the actual name of the method `updateUIForFetching(false)`.

Apart from this, the `InterfaceController` class is familiar territory, and does nothing more than pass method calls and data to and from the `PBLocationManager` class, and is thus a good illustration of the separation of concerns. The controller knows nothing about where the data comes from or how it is formatted; it only needs to be supplied with the right string data. However, you should read through it carefully, making sure that you understand what is going on.

# Interface

Now we must return to Interface Builder to create the user interface.

1. In the project navigator, select the `Interface.storyboard` file, and ensure that the Assistant Editor is open (*Command-Option-Return*).

2. Drag a **Menu** object onto the interface and select the single **Item** in the Organizer, as illustrated here:

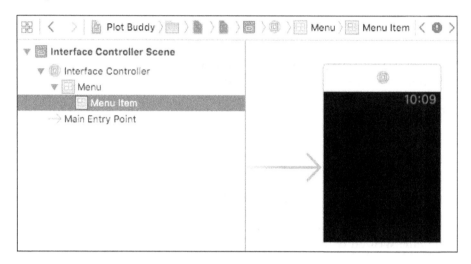

3. In the Attributes Inspector, change the **Title** to **Send Plots** and the **Image** to **Share**.

4. Connect the **Send Plots** menu item to `@IBAction func sendPlotsButtonTapped()` in the source code in the Assistant Editor.

Now we must add the **Group** objects to the interface, given as follows.

1. Drag a **Group** onto the interface and connect to the `@IBOutlet var startStopGroup` in the source code.

2. Set the **Insets** to **Custom**, with a value of **2** for each one.

3. Set the **Radius** to **Custom**, with a value of **8**.

4. Select the **Blue** background color from the color picker (which is the same blue as returned by `UIColor.blueColor()` in code), as shown here:

5. Copy and paste the **Start Stop Group** in the Organizer, and connect the new Group to `@IBOutlet var addPlotGroup` in the code.

6. Set the newly named **Add Plots Group's Vertical Alignment** to **Bottom**.

7. Copy and paste the **Add Plots Group**, and connect this new pasted group to `@IBOutlet var showPlotsGroup` in the code.

8. Add a **Button** object to the **Start Stop Group**, connect it to `@IBOutlet var startStopButton` in the code, change its **Title** to **Start Plotting**, and its background color to **Black**.

9. Add another connection to `@IBAction func startStopButtonPressed` in the code.

10. Add a **Button** object to the **Add Plot Group** with the same property values, except for the **Title**, which should be set to **Add Current Coordinates**, and connect it to to `@IBOutlet var addPlotButton` in the code.

11. Add another connection to `@IBAction func addPlotButtonPressed` in the code.

12. Add a Button object to the **Show Plots Group**, with the same property values, except for the **Title**, which we change to **Show Plots**, and connect it to `@IBOutlet var showPlotsButton` in the code.

13. Add another connection to `@IBAction func showPlotsButtonPressed` in the code.

14. Drag a **Label** object onto the interface, set its **Vertical Alignment** to **Center**, its number of **Lines** to **0**, and its **Font** property to **Footnote**.

15. Connect the label to `@IBOutlet var infoLable` in the code.

Your Interface Builder window should now look like this:

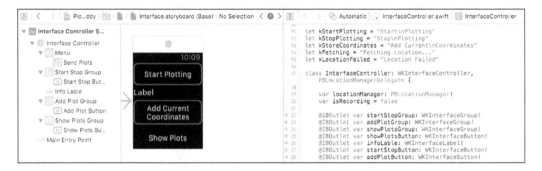

# Test your code

To run the code we simply need to comment out the `sendPlotsButtonTapped` implementation by enclosing it between `/*` and `*/`, like so:

```
@IBAction func sendPlotsButtonTapped() {
/*   WatchConnectivityManager.sharedManager.
sendDataToWatch(locationManager.currentLocations)
*/
    }
```

Now the code will compile, since we no longer have a reference to the `WatchConnectivityManager` that we haven't coded yet.

Hit **Run**, and when your app has launched, tap the **Start Plotting** button. Add a location by tapping **Add Current Coordinates** and your screen will, after a few seconds, look like the illustration here:

If you tap **Stop Plotting**, the **Show Plots** button will become visible, but tapping it will do nothing until we have added the `PlotScene` controller.

We will do this next.

# PlotsSceneInterfaceController

The `PlotSceneInterfaceController` that we are about to create will be responsible for accepting a list of `CLLocation` objects, that is, the `LocationSet` type, and presenting it in a `WKInterfaceTable`. The code here is very straightforward and will present no new challenges.

## CodingPlotsSceneInterfaceController

Create a new Swift file and name it `PlotsSceneInterfaceController.swift`. Replace the `import` statement with the following code:

```
Import WatchKit

Class TableRowController: NSObject {
    @IBOutlet var label: WKInterfaceLabel!
}

Class PlotsSceneInterfaceController: WKInterfaceController {
}
```

We have chosen to define the `TableRowController` class that we will need for out table in the same file as the `PlotsSceneInterfaceController`. We would probably create a dedicated file for this class if its logic were more complex, but in this case we only need to define a single `IBOutlet` property and so we include it here.

Now to add the code for the `PlotsSceneInterfaceController` class.

Add the following code to the class:

```
Class PlotsSceneInterfaceController: WKInterfaceController {

    @IBOutlet var plotsTable: WKInterfaceTable!

    Override func awakeWithContext(context: AnyObject?) {
        super.awakeWithContext(context)
        if let data = context as? LocationSet{ //1
            loadTable(data)
        }
    }
```

```
func loadTable(data: LocationSet) {
    plotsTable.setNumberOfRows(data.count, withRowType:
"TableRowControllerID")
    for (index, location) in data.enumerate() {
        if let row = plotsTable.rowControllerAtIndex(index) as?
TableRowController {
            row.label.setText( //2
                "Location \(index + 1)\n" + //3
                "Lat: \(location.coordinate.latitude)\n" + //4
                "Lon: \(location.coordinate.longitude)"
            )
        }
    }
}
```

The comments in the code are as follows:

1.  We check that the context data is of the correct type and, if it is, we pass it directly to our loadTable method.

2.  The only thing in this code that we have not seen before is the way the text of row.label is spread over several lines for readability. Swift will not allow a single **String** to be defined across more than one line, so it is necessary to use the **+** operator to concatenate the individual strings. However, there is nothing to stop you writing the entire string within one set of quotation marks.

3.  We display to the user the number (i.e. index) of the location in the row, but since most people start counting at 1, we add 1 to the index. There are some interesting articles on the web that argue that children should be taught to start counting at zero. Totally beyond our scope, but worth a search. And if it catches on, we'll come back and update the app!

4.  In this case, we choose only to use the lat and lon properties of the CLLocation object. You may like to investigate adding others once this chapter is complete.

# Creating the UI

Possibly the most error-prone part of using `WKInterfaceTable` objects is hooking them up to the UI in Interface Builder, so, although we're doing nothing new here, we will go through it step by step:

1. Drag an **InterfaceController** object into the Interface Builder window and, in the Identity Inspector, set its **Class** to be **PlotsSceneInterfaceController**.

2. In the Attributes Inspector, set the **Identifier** to **PlotsScene**. If you check back in the `InterfaceController` class code, this is the identifier used in the `showPlotsButtonTapped` method.

 Copying and pasting is the safest way to avoid the frustration caused by typing-errors between Interface Builder's identifiers and the source code, since IB cannot make use of the `String` constants we define in Xcode.

1. Drag a Table object onto the **PlotsScene** interface and connect it to `@IBOut let var plotsTable` in the `PlotsSceneInterfaceController` class's source code.

2. Select **Table Row Controller** in the Organizer pane and set its **Class** to **TableRowController** in the Identity Inspector.

3. In the Attributes Inspector, set **Identifier** to **TableRowControllerID**. You might like to copy and paste this from the `PlotsSceneInterfaceController` class's `loadTable` method.

4. Select the table row controller's **Group** object and set its **Height** property to **Size To Fit Content**.

5. Drag a **Label** object into the table row controller's **Group** object and connect it to `@IBOutlet var label` in the `TableRowController` class's source code.

6. Set the label's Font property to **Caption 2**, its **Min Scale** property to **0.6**, its number of **Lines** to **0**, and its **Width** property to **Relative To Container**.

Sometimes it can seem a little confusing having to hook up all these **Identifier** and **Class** values in Interface Builder, but, like most things, the trick is to take it slowly at first, double checking that you're setting the right IB property with the right name.

# Run the app

Now when you hit **Run**, you will be able to see the second screen displaying all the locations you have added on the first. We're almost done and the only thing the watch app needs is the class that will send data to the iPhone.

# WatchConnectivity

We have done this so often now, we can almost do it blind.

Create a new Swift file, making sure that the WatchKit target is selected, and name it WatchConnectivity.swift. Replace the import statement with the following code, most of which you have seen before:

```
Import WatchConnectivity

class WatchConnectivityManager: NSObject, WCSessionDelegate {

    static let sharedManager = WatchConnectivityManager()

    private override init() {
        super.init()
        if WCSession.isSupported() {
            let session = WCSession.defaultSession()
            session.delegate = self
            session.activateSession()
        }
    }

    func sendDataToWatch(data: AnyObject) {
        guard let contextData = data as? LocationSet else {return}

        let dataArray = contextDataFrom(locationSet: contextData)
        let context = [kApplicationContextDataKey: dataArray]
    do {
            try WCSession.defaultSession().updateApplicationContext(context)
            print("updateApplicationContext succeeded")
        } catch {
            print("updateApplicationContext failed")
        }
    }

    func contextDataFrom(locationSetlocationSet: LocationSet) ->
[[String: String]] {
        var dataArray = [[String: String]]()

        for (index, location) in locationSet.enumerate() {
            let locationDict = [
                "index": "\(index)",
                "lat": "\(location.coordinate.latitude)",
```

```
                    "lon": "\(location.coordinate.longitude)"
                ]
        dataArray += [locationDict]
        }
        return dataArray
    }


}
```

The main thing to note here is that the `CLLocation` data must be formatted as a `Dictionary` object by calling `contextDataFrom(locationSetlocationSet:LocationSet)` in order for it to be sent to the iPhone as Application Context data (we cannot send `CLLocation` objects). We simply iterate through the locations, using the required data from each one to make a dictionary, which we then add to `dataArray` which is declared to be an Array of `[String: String]` `Dictionary` objects. The `dataArray` can then be sent as Application Context data.

If you look carefully through this code, you should now be in the position to recognize everything in it.

# Final test

We now need to uncomment the code of the `InterfaceController` class's `sendPlotsButtonTapped` method, as shown here:

```
    @IBAction func sendPlotsButtonTapped() {
        WatchConnectivityManager.sharedManager.sendDataToWatch(
            locationManager.currentLocations)
    }
```

This enables us to send the data to the iPhone.

The app should now be fulfilling all of the specifications we set out at the beginning of this chapter. And you are hopefully now in the position to include Core Location code in any app that can make use of the data it returns, which is an increasingly large proportion of mobile apps in general.

# Challenges for expansion

If you wish to hone your software skills by extending the core functionality of the Plot Buddy app, here are a couple of ideas to get you started:

- The app uses only two of the `CLLocation`'s properties, but it may be useful to the user to store more of the data returned, such as the time stamp, or the height above sea level. Use the `print()` command to inspect the data in the console and see if you can add more of it to the app.

- At the moment, the app stores only one set of locations at a time and the user must send this data to the phone before recording a new set of locations, or lose it, as it is replaced by a fresh list of locations. With a few tweaks, it is possible to store an array of such location sets and send them all together to the paired iPhone.

- The iPhone does no more than log to the console the arrival of `ApplicationContext` data. Depending on your level of confidence with iOS development, you may wish to display that data on the iPhone screen or package it into an email, or even produce a graphical representation of the data.

# Summary

In this chapter you have learned to incorporate Core Location functionality into your app, which involved adding entries into the app's `Info.plist` file; you have written a custom protocol to facilitate communication between different custom classes and implemented the methods defined in that protocol. Moreover, you have implemented a class's `init` method for the first time—an essential technique in any Swift project—and we have seen how to make method calls more readable by forcing the explicit naming of arguments.

In the next chapter, we will see that there is more to finishing an app than getting the code to run!

# 8
# Images, Animation, and Sound

Now that we have a few functioning apps under our belt, we will take a look at what else there is to do in order to produce a truly release-ready app. One of these topics, the app icon, is mandatory for any project intended for release, while the others are brief explorations of technologies that you are likely to add too many of your own app projects.

The topics covered include:

- Providing a custom icon to an app
- Using an Xcode asset catalog
- Adding an `extension` to a class
- Creating the UI layout in code
- Creating sequential animations
- Adding audio and video playback to your app

## Adding an icon

I don't know about you, but I find that one of the most satisfying points in an app's development is the first time I see its icon appear on the desktop or home screen of whichever platform it runs on. It seems that at that moment the new app is promoted to the status of a real-world application, becoming a member of the exclusive club of useful projects. My hard disk is littered with sketches, try-outs, and projects that for some reason never reached completion and, indeed, my home screens are equally cluttered by the generic app icons added by Xcode to every new project. It's the good ones that get to display their own icon.

So adding an icon is, for me, a sign that an app has progressed far enough that I might be tempted to actually show it to someone.

On the Apple Watch home screen, there is actually a further imperative to add an icon and that is the fact that there are no labels to distinguish one app from another. It's all down to the icon. If you have developed several projects and run them only from within Xcode, whether on the Simulator or on a physical device, this will not have become a problem, but testing an app out in the field, far away from the Xcode's Run button, can quickly become confusing when you have to select one app from several that share the same generic icon on the home screen.

# Icon requirements

Every app submitted to the App Store needs a number of icon images of various sizes for both the iPhone and the Watch. This is mandatory, but the processes of creating and adding these images to the project are not complicated, as you will see shortly.

Although the icons appear circular on the watch's home screen, your files, or assets, as they are called in this context, will be square images and the operating system will take care of applying the circular mask.

Unlike icons on iOS home screens, watchOS icons are not accompanied by text labels, making the design of the icon that much more important. It will be the only thing that your users have to identify your app in the confined space of the watch's screen.

# Technical requirements

Firstly, a few specifications to note:

- The icon must be in .png format
- The image's standard bit depth is 24 bit, but other bit depths are sometimes a good idea (see *Design Considerations* below)
- The image must contain no transparency
- The image must not use a black background, unless it is surrounded by a border, in order to differentiate it from the home screen background

> Failure to adhere to these specifications will very probably mean your App Store submission is rejected, so regard these as hard and fast rules.

Apple's documentation on this topic can be found here:

```
https://developer.apple.com/library/ios/documentation/UserExperience/
Conceptual/MobileHIG/AppIcons.html
```

# Design considerations

In addition to following the image assets technical specifications, there are a few things that we should bear in mind when designing an app icon for the Apple Watch.

At the risk of stating the obvious, there is one facet of the watch that exerts a huge influence on our design considerations: the screen is small and the icons are tiny! It is essential to keep the design simple. Including text in the icon is probably not a good idea and you don't want to be incorporating fine details that will be completely lost in the context of the home screen. Keep it bold and keep it simple.

The icon needs to clearly represent your app. Try to find something that expresses some essential element of your app, or is clearly related to the app's iOS icon on the iPhone.

Apple suggests keeping the colors to those available in the 8-bit palette, which will considerably reduce the size of the image assets (remember, storage on the Apple Watch is not measured in gigabytes). To what extent this is possible for you will depend to a great extent on the colors used outside of the context of your watch app, which may or may not be fixed, such as your website, a pre-existing iPhone app, or a client's corporate identity. Worth a thought, though.

# Using third-party utilities

I have to admit that I am hopelessly lost when faced with a multi-layered, alpha-stripped, gluten-free Adobe Photoshop template. These things are meant for graphics people and the plethora of templates that are available, from both Apple and others, are doubtless a valuable aid to those with the necessary skills to use these excellent tools. But for my own projects, where I don't have a highly experienced — and hugely expensive — designer at my disposal, I prefer to keep things as simple as I can and so we will have a very brief look at some smaller scale alternatives to the industrial might of Adobe.

## Complete icon set

You may wish to download a complete set of Apple Watch icon images for your first go at adding them to you're project. This will allow you to focus on the logistics of getting the assets into the app, while leaving the creative and graphics hurdles for later. You can access a set of icon images (for both the watch and the phone) here:

```
https://github.com/codingTheHole/BuildingAppleWatchProjectsBook
```

## Graphics apps

However, sooner or later you will need to prepare your own graphics. There are doubtless many excellent apps out there and it is beyond the scope of this book to examine a selection of them closely. One such app that I have found to be flexible and simple enough for my own use, and this is a purely subjective recommendation, is called *Graphic* and is available here:

```
http://www.graphic.com
```

It is, however, not freeware, so you may wish to search the App Store for something that is offered for free.

A second utility that I have found most useful is *iConeer* that will prepare all the image sizes you need for any of Apple's platforms, including the iPhone and the Apple Watch, from a single image file. It is available here:

```
http://www.quickgets.com/iconeer/
```

The price of this app was only a couple of dollars and was money well spent, though again, there will certainly be many other apps that also do a fine job, some of them for free.

## Sizes

Whatever software you use, design your app icon using a canvas size of 1024 x 1024 pixels. From the finished graphic you can then create the required image sizes. This large format will then serve you well for all icon-related purposes as well as the larger formats that you may end up submitting to the App Store.

Yes, you may think you're only designing the watch icon, but inspiration is a funny thing and you never know where your creations will end up being deployed.

1024 x 1024 pixels will cover you for all eventualities.

This is not the place to discuss the actual creation of that first graphic, but, once it is finished, you will need to create differently sized versions of it for a number of uses. Most of these will require different sizes for the 38 mm and 42 mm watch faces. In all, you will need the following image sizes (width and height are always identical):

- **Notification Center icon**: 48 and 55 pixels
- **Long-Look notification icon**: 80 and 88 pixels
- **Home Screen icon**: 80 pixels for both watch face sizes
- **Short-Look icon**: 172 and 196 pixels

 These are in addition to the image sizes you need for the iPhone app, which are not covered here, but are included among the icon assets in the GitHub repo above.

## Understanding points and pixels

Xcode expresses image sizes in points, not pixels. A full discussion of the reasons for this is out of the scope of this book, but all you need to know in the case of the Apple Watch screen is that one point translates to four pixels (double the pixels in height, double the pixels in width). Thus, a 24 point image requires an image of 48 x 48 pixels, and this is underlined by Apple's convention of adding the suffix @2x to all image names intended for display on a Retina display (which includes the watch screen).

So icon images will have names such as `icon-24@2x` for a 48 pixel image, `icon-27.5@2x` for a 55 pixel image, and so on, and when we come to Xcode's `xcassets`, we will see the same convention.

One exception to this is the **Apple Watch Companion Settings** image, which must also be provided in @3x resolution (for the iPhone 6+ screen).

# Importing images into the project

We will assume you have a full set of icon assets of the required sizes, which you have completed in a graphics app or downloaded from the GitHub repository listed above.

At the time of writing, a list of required assets is available from Apple here:

```
https://developer.apple.com/watch/human-interface-guidelines/icons-
and-images/
```

The process we must follow is very straightforward; we simply add the files to the project and then assign the appropriate asset sizes in Xcode's asset catalog editor.

Select the `Assets.xcassets` file of the WatchKit App (not the iPhone app) in the project navigator and you will see the asset catalog editor showing empty fields for the required watch icon assets, as illustrated below:

The easiest way to add the icons is by dragging them directly from the finder onto the appropriate empty fields:

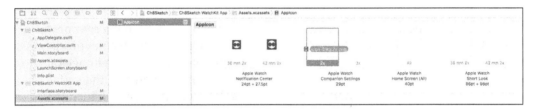

Be sure to select the right image, or you'll get a warning from Xcode:

Once you have dragged in all the necessary image assets, your asset catalog should look like this:

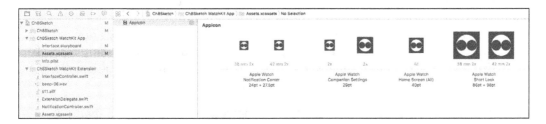

Notice that you did not need to explicitly tell Xcode to copy the image assets into the project folder, this is done automatically, as can be seen in this screenshot from the Xcode project's folder in the Finder:

Launch the watch app in the simulator and, once it is running, type *Control-Shift-H* twice, which will take you to the home screen, where you will see your app represented at last by a custom icon. If you launch the iPhone app (selecting the correct scheme, of course) and open the *Watch* app, you'll see your app listed among the watch apps for inclusion/exclusion on the watch, also displaying the custom icon.

Congratulations, your app now looks as cool as Apple's do.

# Animation

Way back in *Chapter 2*, *Hello Watch*, we used some simple animation to change the background color of the `WKInterfaceGroup`, which, if you remember, involved calling a method available in `WKInterfaceController` which takes two arguments, the duration of the animation and a block of code that contains the changes in appearance that should be animated. We saw that even the subtlest of animations can add a sense of "something happening" on the screen, with only a tiny amount of code. Plenty of bang for the buck there.

We are, however, faced with certain limitations of the methods that WatchKit provides us with, one of which is that all the animated changes take place concurrently; it's all at once or not at all. Yet we can easily imagine that even the most modest of animations (we're really not talking Pixar here) might require a series of discrete steps in a particular order. We might want to animate a change in the position of the WKInterfaceGroup and then have it change its background color, for example. Or we may wish to set up a chain of small views that appear and disappear sequentially to create an activity indicator.

In order to achieve effects such as these, we will need to write our own method, extending those provided by watchOS, and then call this method to realize our sequential animations.

# Creating AnimationInterfaceController

What we need to do, then, is create a method similar to WKInterfaceController class's animateWithDuration method that takes an additional argument, which will be a block of code that will be called once the animation is finished. That code could be anything at all, including a further call to animateWithDuration. Thus, we would now have two separate animations chained together.

## Create a new project

To do this, first create a new watchOS project, as we have done previously. For the sake of simplicity, deselect the **Notification** and **Glance** scenes when setting up the project. Give the project any name you like, as we will not be producing a complete app here; we are more concerned with extending our development skills in an area applicable to just about any app we create.

Let's rename the InterfaceController class:

1. Select the InterfaceController.swift file in the project navigator. In the code, change the name of the InterfaceController class to something a little more specific, AnimationInterfaceController, as shown here:

```
import WatchKit
import Foundation

class AnimationInterfaceController: WKInterfaceController {

    ...
```

2. With the `Interface.storyboard` file selected, select the `InterfaceController` object and use the Identity Inspector to change its class to `AnimationInterfaceController`.

# Extending AnimationInterfaceController

We will now add an `extension` to our `AnimationInterfaceController` class. Replace the code in the file with the following (you'll notice some code that we do not need from the template has been removed, though leaving it there would have no negative effects):

```
import WatchKit

class AnimationInterfaceController: WKInterfaceController {

    override func awakeWithContext(context: AnyObject?) {
        super.awakeWithContext(context)
    }
}

extension AnimationInterfaceController //1
{
    func animateWithDuration( //2
        duration: NSTimeInterval,
        animations: () -> Void,
        completion: () -> Void) //
    {

        self.animateWithDuration(duration, animations: //3 animations)

        let delay = Int64(duration) //4
        let time = //5
        dispatch_time(DISPATCH_TIME_NOW, delay * Int64(NSEC_PER_SEC))

        dispatch_after( //6
            time,
            dispatch_get_main_queue(),
            completion)
    }
}
```

The comments in the code are as follows:

1. This extension will apply to the `AnimationInterfaceController` class (and any of its subclasses, should we need to create any). We could have extended the `WKInterfaceController` itself, since `AnimationInterfaceController` is declared to be a subclass of that, but we're assuming that only `AnimationInterfaceController` instances need the new animation functionality.

2. We declare a method that takes three arguments: the animation's duration, a block of code that will comprise the animation itself, and a further block of code that is to run once the animation is complete.

> Because the method signature includes the arguments list (the stuff in brackets after the method name), there is no confusion with the `WKInterfaceController` class's `animateWithDuration` method, which takes fewer arguments.

3. When the method is called, we first run the animation block supplied in the second argument by passing it to `WKInterfaceController` class's own `animateWithDuration` method.

4. That animation will now begin on a separate thread, while our code continues to execute. What we do now is convert the `duration` argument into an `Int64` (named `delay`) which is the type we'll need to create a delay in code execution.

5. We use that delay to create a `dispatch_time` which is basically the time at which we want our `completion` code to be executed.

6. We use the operating system's threading framework (often referred to as Grand Central Dispatch) to run the `completion` code on its `main_queue` after `time` has elapsed. Think of this as a way to create a countdown to the execution of a block of code.

> The finer points of threads and concurrent code execution are complicated and beyond the scope of this book. Don't worry if this stuff looks intimidating, it's an advanced topic and the best thing to do here is simply accept the code as is and look forward to some future date when it will all seem very simple.
>
> There's not much that scares developers more than concurrent programming, so you're not alone.

We will now have this method at our disposal in all instances of our `AnimationInterfaceController` class. The `completion` block of code will start at the same moment that the `animation` block finishes.

## Add Outlets to AnimationInterfaceController

Add the following outlets to the `AnimationInterfaceController` class:

```
classAnimationInterfaceController: WKInterfaceController {

    @IBOutlet var label: WKInterfaceLabel!
    @IBOutlet var outerGroup: WKInterfaceGroup!
    @IBOutlet var buttonAGroup: WKInterfaceGroup!
    @IBOutlet var buttonBGroup: WKInterfaceGroup!
    @IBOutlet var buttonA: WKInterfaceButton!
    @IBOutlet var buttonB: WKInterfaceButton!

    @IBAction func buttonATapped() {
    }
    @IBAction func buttonBTapped() {
    }

. . .
```

We will then be able to hook up our UI elements from Interface Builder

# Creating the UI

We will take a slightly different approach to creating the user interface, which this time will involve doing much less of the layout work in Interface Builder and instead do it in code.

How the layout of an interface is divided between Interface Builder and runtime code is partly a matter of which is more suitable, but also a question of a developer's personal preference. Some developers are more comfortable doing as much as possible graphically, using Interface Builder, and the advantages here are clear enough – you get to see immediately the results of changes you make.

In the case of WatchKit, there are also considerable limitations as to how much you can do in code: you cannot, for example, add UI elements dynamically at run time, it all needs to be created in advance. There are also some object properties that cannot be set in code, for example the `WKInterfaceLabel` number of lines property.

However, when it comes to laying out the UI, many developers prefer to do as much as possible in code (disclosure: this author is one of them) and, if you prefer to picture the appearance of the screen in your head and then translate that into code, you'll find that there are many benefits to doing so. One of these is that you can work at the code level uninterrupted by frequently referring back to Interface Builder, instead much of the layout takes place in code early on in the view's lifecycle (we will use the awakeWithContext method) and provides a clear overview of the values of UI element properties when a view launches.

Since the essence of our animation code will be changing these properties over time, we will find it simpler to follow what's going on if we do as much of the layout as possible in code.

All we need to do in Interface Builder is drag the various UI elements we need onto the **Animation Interface Controller** or, even easier, drag them directly onto the document Outline.

Follow these steps to add the UI objects we will configure in code later:

1. Start by adding a **Group** object:

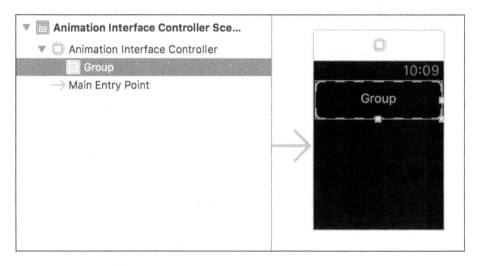

2. Complete the hierarchy of **Group** and **Button** objects as illustrated here:

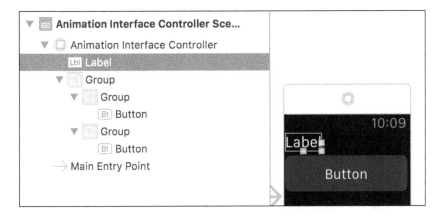

3. Hook up each of the objects to the appropriately named IBOutlet and IBAction code in the source file, so that the objects in the document outline are assigned the appropriate names, as shown below:

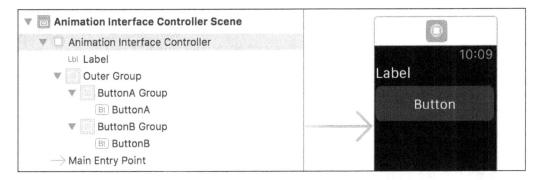

And that's it for our work in Interface Builder. No need to set any properties of the individual objects, which will all be done in code.

# Adding code to AnimationInterfaceController

Now we must complete the code of our `AnimationInterfaceController` class. We will declare a few `UIColor` constants, lay out the UI, and finally add the animation code using the new functionality added by the extension to the class.

## Some Constants

After the import statement (but outside of the class's code), add the following declarations:

```
import WatchKit

let color1 = UIColor.lightGrayColor()
let color2 = UIColor.blueColor()
let color3 = UIColor.redColor()

class AnimationInterfaceController: WKInterfaceController {

    . . .
```

This will make our code tidier, allowing us to refer to the colors that we want instead of repeatedly writing the code we use to obtain them. In a large project, with many of these constants, we would likely place them in a file of their own, thus separating further the use of our colors from their implementation.

These are very good habits to get into. Really.

## Setting the UI layout in code

Now we add a function that will do everything we need to set out the layout and behaviors of the UI objects we created in Interface Builder. You will notice that all of this code is equivalent to setting the properties in IB.

Add the following code to the `AnimationInterfaceController` class:

```
func layoutUI()
{
    label.setText("Choose One")
    label.setAlpha(0.0)
    label.setHorizontalAlignment(.Center)
    label.setVerticalAlignment(.Center)

    outerGroup.setHorizontalAlignment(.Center)
    outerGroup.setVerticalAlignment(.Top)
```

```
        outerGroup.setWidth(80)
        outerGroup.setHeight(20)

    func setUpButtonGroup(group: WKInterfaceGroup) {
        group.setBackgroundColor(color1)
        group.setCornerRadius(8)
        group.setRelativeWidth(0.1, withAdjustment: 0)
        group.setRelativeHeight(1.0, withAdjustment: 0)
        group.setContentInset(
            UIEdgeInsets(top: 4, left: 2.0, bottom: 4, right: 2.0)
    )
    }


        setUpButtonGroup(buttonAGroup)
        setUpButtonGroup(buttonBGroup)
        buttonAGroup.setHorizontalAlignment(.Left)
        buttonBGroup.setHorizontalAlignment(.Right)

        buttonA.setTitle("")
        buttonB.setTitle("")
        buttonA.setBackgroundColor(UIColor.blackColor())
        buttonB.setBackgroundColor(UIColor.blackColor())
        buttonA.setVerticalAlignment(.Center)
        buttonB.setVerticalAlignment(.Center)
        buttonA.setHorizontalAlignment(.Center)
        buttonB.setHorizontalAlignment(.Center)


    }
```

You might be surprised to see a function defined inside another function, as is the case with the setUpButtonGroup method.

The idea here is that both of the buttonGroup objects will have most properties set to the same values and, rather than write the code once for each button, it is more succinct to pass each buttonGroup into a method that sets the properties. However, that method has no use outside the layoutUI function and so we place its definition inside the method that uses it.

## Run the code

Let's see if this has all worked out the way we want it to. Add the following line of code to the `awakeWithContext` method:

```swift
override func awakeWithContext(context: AnyObject?) {
    super.awakeWithContext(context)
    layoutUI()
}
```

Now when we hit **Run**, we should see the app launch and present us with the following screen:

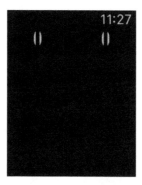

## Tweak the code

You may be tempted to tweak a few of the property values that we have set in the `layoutUI` method. Go ahead, it's your app, after all.

Bear in mind that this code is the result of hours of such tweaks and experiments and that few developers would be able to accurately imagine what the interface would look like just from reading that code, at least, not without investing considerable time.

Play around with it and try to work out which properties are contributing what to the appearance of the UI as it stands. That should provide a few hours worth of geeky entertainment, if nothing else.

Having done that, it's time to code some actual animation!

# Completing the animation code

Add the following two method declarations to the `AnimationInterfaceController` class:

```
func startAnimations(){
}

func startSecondaryAnimation(){
}
```

The first of these methods, `startAnimations`, will use our extension of `AnimationInterfaceController` to run its own animations and, on completion, call `startSecondaryAnimation`, which will, naturally enough, contain the continuance of our sequential animation.

We will call `startAnimations` from the `WKInterfaceController` class's `awakeWithContext` method, so add the following line of code to that:

```
override func awakeWithContext(context: AnyObject?) {
    super.awakeWithContext(context)
    layoutUI()
    startAnimations()
}
```

Next, we'll add the code for the first animation to the `startAnimations` method:

```
func startAnimations(){
    outerGroup.setAlpha(0) //1

    animateWithDuration(3.0, //2
        animations: {
            self.outerGroup.setAlpha(1.0) //3
        },
        completion: startSecondaryAnimation) //4
}
```

The comments in the code are as follows:

1. We make the label invisible, before we start to fade it in.
2. We call the method with which we extended `AnimationInterfaceController`, setting a duration of `3.0` seconds.
3. The property that we want animated is the `label` class's `alpha` property, which gets increased to a value of `1.0`, which represents full opacity.

4. We pass in the `completion` block of code as the third argument. This time, instead of the argument being in the form of a closure, that is {//some code here}, we pass it a complete method, `startSecondaryAnimation`.

 Note that we don't append parentheses; we are not calling the function, but passing it as an argument to another function (which will call it). This allows us to define the implementation of the completion code separately, elsewhere in the code.

If you run the code now, you will see the gray lines fade in over a duration of 3.0 seconds.

Now we add the code for the `startSecondaryAnimation` method:

```
func startSecondaryAnimation(){

    animateWithDuration(5.0, //1
        animations: { //2
            self.outerGroup.setVerticalAlignment(.Bottom)
            self.outerGroup.setRelativeWidth(
                1.0, withAdjustment: 0)
            self.outerGroup.setHeight(40)

            self.buttonAGroup.setBackgroundColor(color2)
            self.buttonBGroup.setBackgroundColor(color3)
            self.buttonAGroup.setRelativeWidth(
                0.45, withAdjustment: 0)
            self.buttonBGroup.setRelativeWidth
                (0.45, withAdjustment: 0)
        },
        completion: {
            self.animateWithDuration(2.0, //3
                animations: {
                    self.label.setAlpha(1.0)
                },
                completion: {
                    self.buttonA.setTitle("blue")
                    self.buttonB.setTitle("red")
            })
        }
    )
}
```

The comments in the code are as follows:

1. Once again, we use the method we defined in the extension to `AnimationInterfaceController`.

2. The `animations` argument is passed as a closure, although we could have have implemented this code in a separate function and passed that function as the argument, as we did in `startAnimations`, above. Because we are passing a closure, we need to prepend the references to the UI objects with `self`.

3. The code in the completion closure calls, you guessed it, our extension method. Nesting like this means that we could continue chaining a sequence of animations of any length we like. This closure first fades in the `label` and then the `completion` block sets the `Title` properties of the buttons from the empty string to **blue** and **red**.

You can probably tell from reading the code that the `outerGroup` element is going to change size and position and that the `buttonGroup` objects are going to change color and size.

Give it a go. When you hit **Run**, you should see the gray lines fade in for 3 seconds, then move downwards and outwards while taking on a blue or red color and starting to look more like the outlines of buttons that they actually are. Finally, after 5 seconds, the titles of the buttons are shown and the user is invited to choose one of them.

Well, we already have the buttons connected to `IBAction`, so all there is left to do for this example is to add some code to them.

If the user chooses **blue**, she is choosing not to enter The Game (for the time being in a metaphorical sense).

Add the following code to the `buttonATapped` method:

```
@IBAction func buttonATapped() {
    startFinalAnimation()
}
```

Then add the `startFinalAnimation` method to the `AnimationInterfaceController` class:

```
func startFinalAnimation()
{
    self.buttonA.setTitle("")
    self.buttonB.setTitle("")

    animateWithDuration(5.0,
```

```
        animations: {
            self.outerGroup.setAlpha(0)
            self.label.setAlpha(0)
        },
        completion: {
            self.label.setText("Your Choice")
            self.animateWithDuration(1.0,
                animations: {
                    self.label.setAlpha(1.0)
                }
            )
        })
}
```

For one final time, we call our extension method, which, as you can probably tell, fades out the buttons (that are in the outerGroup) and the label and then fades the label in with a new text.

If the user chooses **red**, well, I'll leave it up to you to extend the rewards of the red button beyond this:

```
@IBAction func buttonBTapped() {
    print("Entering the Matrix?")
}
```

# Less is more

Please forgive the sermon, but go easy on animations (unless you are developing a game, or an app in which animations are a core functionality). If you know anybody with gray hair and a taste for 80's pop music, ask them how much they enjoyed flashing HTML in the early days of the internet.

Used with taste, animations can add some real value to your app, but overdo it and your users may quickly become tired of having to wait for a button to do a lap of honor around your interface before they can continue using the app.

Here ends the sermon.

# Run the app

All being well, you should now see the UI fading in, moving around, and on selection of the **blue** button, fading back out again.

You did connect up the UI elements with the source code, didn't you?

Now, about that **red** button.

# Audio and Video

Okay, we've got the visuals, let's make some noise!

It has to be said that audio functionality on the Apple Watch is very limited. It is yet to be seen how the platform develops in this respect, but, as things stand, there is not a great deal you can do with media at the moment beyond simply playing an audio or video file. Perhaps the rationale behind this is that a device that beeps and squeaks at every opportunity is likely to be an irritation not just to the user who is wearing it, but to everybody else in the near vicinity as well. Everything that was said about animation applies to audio tenfold and is probably considered less of a *must-have* than an *if-you-must*. And, of course, there is no headphone socket to relieve your family members of this new source of background noise.

However, that said, if we keep an open mind on the subject, we do have an opportunity to explore the extent to which we can use audio on the Apple Watch, and it is likely that there will be a hundred hitherto unimagined use cases just waiting to become as mainstream as puppy videos on an iPhone are today (and believe me, nobody was thinking of that in the 1990's).

There are a few approaches to getting audio to play on the watch and we will employ the simplest here. This involves employing the `WKMediaPlayerController`, which is provided by WatchKit. This player is also capable of playing video, so although we will be using a very small audio file in this example, there is no additional code necessary to play an **.mp4** video file.

There is a small audio file available in the GitHub repo for this book if you don't have any media to hand:

`https://github.com/codingTheHole/`
`BuildingAppleWatchProjectsBook`

We will use it for this chapter's example, since such a small file will load very quickly and, further, will not bloat the size of the app.

# Adding a media file

When we say media file, we are referring to any of the formats that can be played by the `WKMediaPlayerController`, which include the `.aiff` format we will use here and the `.mp4` video format.

To add a media file to your project, simply drag it from the finder into the project navigator and select **Copy items if needed**, ensuring that the **WatchKit Extension** target is selected, as illustrated here:

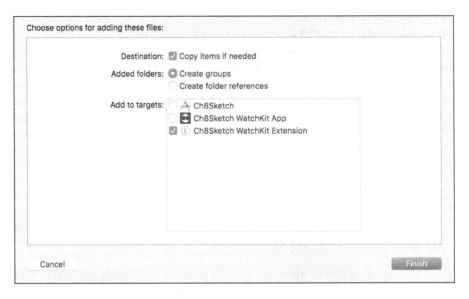

# Adding the code

Now that we have added the media file to the WatchKit Extension's bundle, let's write a function that we can call from anywhere in our code to present the system's own media player, loaded with that file.

Add the following code to whichever `InterfaceController` object is required to present the player. If you still have our `AnimationInterfaceController` class open, you could use that:

```
func playSound() {
    if let assetUrl = NSBundle.mainBundle().URLForResource( //1
        "donk", withExtension: "aiff")
    {
        let options = //2
        [WKMediaPlayerControllerOptionsAutoplayKey :"true"]
        presentMediaPlayerControllerWithURL( //3
            assetUrl,
            options: options,
            completion: { didPlayToEnd, endTime, error in //4
                // completion code here, if any
        })
    }
}
```

The comments in the code are as follows:

1. We get the URL of the asset we want to play, which resides in the app's `mainBundle`. Careful with the typing, any mistakes here will mean that the asset (that is, the file) doesn't load, but there will be no indication that an error has occurred.

2. We create a dictionary of options, which in our case sets the `WKMediaPlayerControllerOptionsAutoplayKey` to `true`. As you can imagine, this means that the file will play as soon as it is loaded.

3. We now make the call to `presentMediaPlayerControllerWithURL`, passing it the `assetUrl` and `options` we just created and a `completion` closure, which the system will call, supplying a `Bool` value that tells us whether or not the file played to the end, an `NSTimeInterval` informing us how much of the media was played, and an `NSError` if anything went wrong.

You must now call this method from somewhere in your code. At the moment we have a red button that doesn't do much, so let's add the call to its `IBAction`, `buttonBTapped`:

```
@IBAction func buttonBTapped() {
        print("Entering the Matrix?")
        playSound()
    }
```

Now when you select **red** from the two buttons, you will not only see the enigmatic `Entering the Matrix?` message in the console, you will also be presented with the watchOS media player, containing possibly the most inappropriate sound ever to accompany an initiation into the dark and mysterious world of Morpheus.

You might like to replace the media player with an exciting game instead.

# Summary

In this chapter, you have learned how to use an Xcode asset catalog to add the required icon images to your watch app and you have used an extension to provide advanced animation functionality to an `InterfaceController` object beyond that supplied by WatchKit itself. You have learned to lay out the user interface in code as an alternative to using Interface Builder and you have seen how to add media playback to your apps.

Although the techniques presented in this chapter will already elevate your projects to an altogether higher level, giving you opportunities to further engage and delight your users, they are intended as first steps toward a rich and detailed understanding of the wealth of opportunity offered by the Apple Watch. The ways in which we will all be using these features in the not-too-distant future are largely in the hands of you, the developers, whose task it is to explore this wide open field of mobile technology.

In the next chapter, we will look at what needs to be done before submitting your app to the App Store for approval by Apple.

# 9
# Wear It, Test It, Tweak It, Ship It

Although there is more than a modicum of truth in the saying *Software development is a process, not a product*, there comes a point when the current version of an app has become a release candidate and a number of post-development housekeeping tasks remain to be completed before said app is ready for shipping.

This chapter deals broadly with two areas—testing the app once it is more or less ready for release and submission of the app to the Apple App Store.

These two areas have something in common, apart from being essential parts of the app development process—both tend to be underestimated in terms of the time they take. You are unlikely at this stage to be working to a strict release schedule, it is true, but often there is a temptation to rush things in this final stretch, now that the code is written and you are apparently (pardon the pun) ready to go, and the tasks that get overlooked or hastily completed will come back to bite the impatient developer.

These are the final steps you'll take before you hit the upload button, after which it's all down to Apple, or most of it anyway. And once your app is accepted by Apple (and this chapter will do its best to help you achieve that lofty status), it's out there in the wild, fending for itself. However deep and genuine our respect and affection for our users may be, the fact is that ten thousand of them (or even just ten, for that matter) represent the greatest challenge to you as a developer and to your app.

In this chapter we will cover the following topics:

- Installation on a physical device
- Testing in the field

- Xcode distribution settings
- Requirements for iTunes Connect
- The App Store submission and review process

# Installation on a physical device

You may already have an Apple Watch and you may have been using it to build and run the apps presented in this book, all with no problems at all. And if you have, good on you, you were in luck.

Quite possibly you have a watch, and you have made a detour here because things aren't running quite as smoothly as we would wish. If that's the case, we will soon have you up and running.

# What if you don't have an Apple Watch?

It is also completely plausible that you have read this far, using only the Xcode's Watch Simulator app and Xcode itself to test your code.

By now you'll have had considerable opportunity to decide whether or not you're going to get serious about developing for the Apple Watch. This is handy, because you really do need to commit to the purchase of a physical device to go any further in the process of turning an idea in your head into an app on somebody's wrist.

Although you can test at least 90% of your app's code on the Simulator, there are many aspects of an app's behavior on a real watch that will differ significantly to that on the Simulator. The latter is part of a huge computer operating system, with more speed, memory, and storage space than an Apple Watch could dream of (or is likely to possess for a decade or so, though who can tell?). It has access to an internet connection that is several orders of magnitude more reliable, and probably a good deal faster than the network conditions that your apps will have to deal with (safely and gracefully) in everyday use. You are unlikely to have walked very far with your MacBook in your hands, testing the location code, am I right?

There are just too many things that could be seriously amiss with an app to release it to the public at large without having done extensive testing on a real, shiny, visually fascinating, haptically vibrating, physical Apple Watch.

# Registering your device

This process is the same as you'll probably have done already for your iPhone. Fortunately, it is a process that has become less and less obtrusive as the iOS platform and its development practices have matured, and it is now just a matter of telling Xcode that you will be using this device for development purposes.

The one prerequisite for device registration, apart from the watch itself, of course, is current membership of Apple's Developer Program. This isn't free, but at US$99 a year, it will sting a lot less than the cost of the watch. If you feel that paying Apple to develop the apps that make their platform so popular is a little unfair, remember that this small but significant financial hurdle contributes to making sure that apps in the App Store are being written by people who are reasonably serious about it, in contrast to some other platforms on which the user must wade through oceans of dross to find the apps worth serious consideration.

The fine details of setting up your membership are beyond the scope of this book, but the following link will set you on the right path:

```
https://developer.apple.com/support/compare-memberships
```

# Pair your Apple Watch and iPhone

You must pair the watch with an iPhone to be able to use it. It's a fun process involving pointing the iPhone camera at the watch while it displays a curiously pleasing particle-cloud animation that contains whatever hidden secrets the iPhone is looking for to complete the pairing. Sometimes Apple's design team pulls a rabbit out of the most unexpected of hats.

For the full details of pairing, refer to the instructions that come with it.

# Select the device in Xcode

When you plug your physical iPhone, give Xcode the time it needs to perform some administrative duties, as indicated in the illustration below:

Once Xcode has finished indexing the files it needs to, select the WatchKit App scheme and then select your iPhone and Apple Watch as shown here:

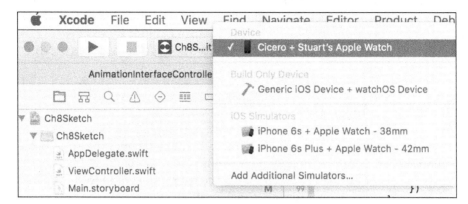

 Make sure you have the WatchKit App scheme selected, if you don't then Xcode will only allow you to select the phone and not the phone/watch combination that we need here.

If you run into the problem illustrated below, you'll need to give Xcode a little more time to work out that the iPhone is indeed paired with a watch:

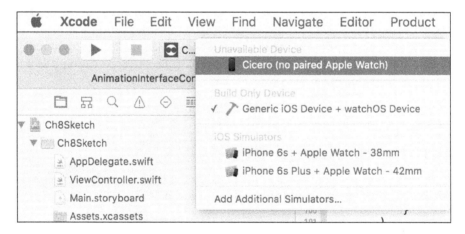

You may also run into this if you are using a beta release of Xcode (if you don't know what that is, then you're not using one). The reader is encouraged, as strongly as is possible without bold text and exclamation marks, not to use a beta release while getting familiar with both the hardware and software aspects of developing for the Apple Watch. The betas throw up too many surprises. You don't want surprises.

# Running on the device

I have found that it is a good idea to run the iPhone app scheme first, which will confirm that the companion iOS app is running healthily.

Now, with the **WatchKit App** scheme selected, hit **Run**.

Be patient, installation can take a little while, particularly the first time you install an app on the watch (all sorts of stuff gets copied across with the first build), or even the Simulator. When this is taking place on a physical device, it needs even longer. It has often taken a few minutes for this first installation to complete so keep your eye on Xcode, which will inform you that the app is being installed on the watch, as can be seen here (this is just a selection):

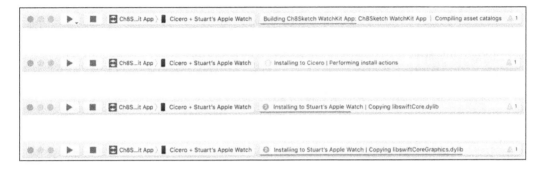

And so it goes on, seemingly forever. As you can see, there is a large range of data that need to be copied firstly to the iPhone and then from there to the watch.

Sooner or later, though, your WatchKit app will be installed on your Apple Watch, where you can test it rigorously under out-in-the-wild conditions.

# Installation troubleshooting

This section deals with tackling installation failures. Most of the solutions offered here apply equally to the physical Apple Watch and the Simulator app.

Without doubt, the most frustrating thing that we are faced with as developers is hitting the Run button, watching the app build successfully, only to be faced with some incomprehensible installation failure. When this happens to you, and it will happen sooner or later, rest assured that you are in good company, it happens to everyone.

Sometimes it will be something that you have done wrong that you will find and fix, sometimes it will be something that inexplicably disappears after a re-launch and remains forever a mystery (and you'll never know what you did that fixed it), and sometimes it will be all Xcode's fault and you will find yourself completely blameless but equally frustrated. It's not something that ever goes away, but it is something that you come to accept as a part of the job. If you need a silver lining, bare in mind that with each installation failure, your stress levels will spike slightly less than the time before.

In the immortal words of Douglas Adams:

> *Don't Panic!*

# Restarting, really?

Yes, really. It does indeed feel like a solution that no longer has a place in the technologically advanced world that we inhabit as software engineers, but it's very often the solution that works. The question is restart what, exactly?

So let's take things in the order in which a restart (or a deletion) is likely to be least intrusive:

## Restart just the app

This is a nice quick one, no beads of sweat on your brow here. Whether on the Simulator or on a physical device, the procedure is the same: kill the app by hitting the Stop button, or simply type Command- (that's *Command-Period*, or *-Full Stop*, depending on which flavor of English you speak).

In the case of developing for the Apple Watch, this has proven to be a fix for at least half of the installation failures that arise. We are dealing with a new platform and there are many new challenges not only to ourselves, but also to those esteemed colleagues that are developing the watchOS platform and Xcode itself, not least of which is the fact that a single platform, iOS, installed on one device (or Simulator app), is responsible for the installation of another app, on its own platform, watchOS, on a different device (or Simulator).

When you get a watch installation failure message pop up in front of you, just rerun the app before you even bother to sigh. Happens all the time.

## Restart the Simulator

This frequently flushes out whatever problem Xcode or the Simulators were having.

## Delete the app from the Watch Simulator

This helps too. But how do you delete an app from the Watch or the Simulator?

If you have the Watch Simulator running, type *Command-Shift-H* to return to the watch face if you don't see it already and then the same again to get to the home screen. If you are running on a physical device, press the Digital Crown instead of *Command-Shift-H*.

Locate your app (aren't you glad you added those custom icons now?) and *Click-hold* on it in the Simulator, or *Tap-hold* on it on the watch until you see the deletion **x** appear, as illustrated below:

This will doubtless look familiar to you from the iPhone home screen which appears very similar. Tap or click on the **x** to delete the app and any data that it has stored on the device. You'll be asked to confirm the deletion by the system.

The next time you hit **Run** in Xcode, the app will be installed afresh and very often the problem (whatever it was) is solved.

## Delete the app from the iPhone Simulator

You may find that that it helps to delete the iOS app from the iPhone, which will also delete the watchOS app on the watch or Simulator.

## Delete the Derived Data directory

Welcome to the iOS development community's favorite fix. Tucked away deep in your file system is a folder called `DerivedData`, where Xcode stores a lot of data that it uses to build your app but doesn't want to have to calculate anew with every build. Sometimes that data causes problems, ones that can be simply solved by deleting the folder containing it. This is one of the most popular pieces of advice you'll get from experienced developers who try to help on various internet forums (see below), and it's as easy to do as it is safe.

You no longer need to drill down through your file structure to get to this data. Xcode provides a button to delete the data in the **Window | Projects** window, which lists all the derived data folders, as shown here:

Don't worry about deleting this data, it will be replaced during the next build process (which will take fractionally longer).

## Reload the Xcode project

This has worked often enough that it's worth inclusion in this list. Just close the project and whatever mess Xcode has got itself into disappears with it. When you restart the project, you can continue working.

## Restart Xcode

Again, this can solve some of those hidden issues that are none of your doing. Xcode has a monstrously large amount of work to do every time you nonchalantly hit the **Run** button (that's why it takes so long) and sometimes things get in a twist. Rebooting Xcode can untwist some of them.

## Restart Xcode

This is unlikely to help. Try everything else first. Then try everything else again.

# Read the error messages

When things go wrong, Xcode tries to help. Very often, Xcode believes the most helpful thing for you is to be told that `Error 2554` has occurred.

Don't be intimidated by this, hardly anyone knows what these error codes mean. But do make a note of them (a screenshot is the fastest way I know of recording the errors for future reference), you'll need them when you turn to your colleagues on the web.

It is quite hard, especially when beginning development, to keep a cool head when things fall apart and making sense of these error messages seems an impossibly difficult task. It is tempting to assume that the rest of the world knows exactly what they mean and a feeling of hopeless inadequacy is added to the problem. Don't succumb to that temptation; even the most advanced developer stands clueless before one of these errors on a regular basis.

## Help online

Don't be afraid to ask for help when errors occur. Some of the most valuable resources available to developers of all experience levels are the forums where like-minded people try to assist and learn from each other.

### Stack Overflow

This is perhaps the number one go-to when you need help. S.O. has more than a million registered users, some of whom will have run into just about any problem that you can imagine:

```
http://stackoverflow.com
```

To ask a question, you will need to be a member of the forum, but membership is free and takes about a minute to get done.

### Apple developer forums

Apple's developer forums are a similarly helpful bunch and unlike Stack Overflow, they're all working on Apple platforms. Access is available to anyone with Developer Program membership:

```
https://devforums.apple.com/index.jspa
```

As with any technical forum, be specific, be polite, include the solutions you have already tried, and try to include any error messages that you think will help people understand your issue.

# Testing in the field

Once you have your app running on your device, you begin to gain an insight into exactly how your app will feel and behave for your users, an insight much deeper than that which is possible using the Watch Simulator app on a desktop computer or laptop.

# Wear it all day

Spend a significant amount of time getting familiar with both your own app(s) and those of other developers, big and small. You need to develop an intuitive feel for users' expectations while using the Apple Watch (although it is entirely up to you whether you choose to fulfil those expectations or delight the user with something new). You may well find that firmly held opinions about what should happen and how can change significantly after a period of hours, or weeks, wearing the watch and using it to engage with your app.

Are your app's implementations of common use cases different from other apps? If they are different, are they better? Are they really better?

 Don't get too hung up on following the pack. If you genuinely believe that you have found something better to offer your users, don't let the nay-sayers tell you any different. On the other hand, don't forget to listen to user feedback. When people don't use your app, it sends a clear message, one that you'd be wise to catch early.

The value of spending some quality time with your app should not be underestimated—this really is an essential stage of development. The app needs to be beyond a certain level of completion to be usable, so this period of testing will inevitably lead to revisiting design and implementation decisions you made some time ago.

Be ready to tweak.

# Scenarios not to be forgotten

Pay particular attention to those scenarios that you can't replicate during development at your desk or anywhere you may set down your laptop (I prefer the beach, personally). Test how your app reacts to your walking outside of a wifi zone; test what happens when the app loses contact with its iOS companion and check if the app reacts to your walking back into range as you intended it to.

How does your app feel after a few days of use (presumably not continuous!)? Is the UI performing to expectations? Are the individual elements easily reachable? Does everything feel solid and 'snappy' enough? Is there anything that you feel is unnecessary and could be removed to make room for other features?

Of course, sooner or later, you will have had enough of asking yourself these questions and pine after a second opinion.

# Beta testers

If you've not come across the term before, a beta tester is someone who tests a version of the app that is intended for release. It is not uncommon that you make a few unexpected and (usually unwelcome) discoveries once someone else starts to use the results of all that hard work.

If you can, try to find someone that will test your app for you. This could be difficult at the moment, since the Apple Watch is a new platform and not many people own one. But it's really worth doing, if you can. At the very least, hand your watch to a trusted friend or colleague and observe how they interact with your app.

Try to get a clear picture of the following points:

- Does the tester understand what the app is supposed to do?
- Does she know how to use the app?
- Is the navigation clear and can the user easily navigate back to the app's main screen?
- Do all of the features work as you expected?
- Do all of the features work as the tester expected?

And, whatever you do, don't take any forthcoming criticism personally. Tough though it is to hear that you haven't quite got it right, remember that this is your opportunity to fix things before your app reaches the broader public, the vast majority of whom will not take the time to give you any feedback at all, whether good or bad.

Large software projects will frequently allocate a quarter of the development time to beta testing and the accompanying fixes and tweaks, which is a good indication of how valuable this phase is.

# Iterate testing and tweaking

Fix what's not right. Clarify what's not clear. Then hand it back to the testers.

Repeat.

# When testing is done

Of course, you could test forever. It's also unlikely that you'll catch every single bug in your code. But at some point you will decide that it's time to release the app into the wild. Everything your app needs has been implemented and everything you have implemented seems to be working satisfactorily.

[
The app doesn't need every feature you intend it to have in the
long run, but you do need a coherent core feature set, the so-called
*minimum viable product* (generally abbreviated in corporate circles
to MVP, mostly because corporate circles love abbreviations).
In addition to that core functionality, most projects will include
a number of other features whose implementations were
straightforward enough to include in the initial release.
]

If you're sure the app is ready, the testers agree the app is ready, and you've checked
all you can check, you're into the final stretch; the preparation of your app for its
submission to the App Store and the handover to Apple's review process.

# Before you submit

Before moving onto submission, check the following points:

- The version and build numbers need to be the same on the target settings for
  the **WatchKit App**, the **WatchKit Extension**, and the iPhone's **App**. If you
  have written both apps from scratch, this won't be an issue, but if you are
  adding a WatchKit app to a previously released iPhone app, you'll need to
  bear this in mind.

- Remove, disable, or comment out all logging to the console.

- Any compiler warning messages should to be resolved. They won't prevent
  you from making a submission, but they are there for a reason, and now is a
  good time to make sure you are not making problems for yourself later on.

# App Store submission process overview

The complete submission process for an iPhone App is beyond the scope of this
book. There are many good resource on the internet that will guide you through
this if you have never done it before, and this chapter will assume that you have
all the information and resources needed for the iPhone app itself in order to focus
on those additional preparations that are necessitated by the addition of an Apple
Watch App to the main companion app.

We will then present a very succinct summary of the processes involved in preparing
your app for upload to the App Store, once again concentrating on the WatchKit
app's requirements beyond those of a purely iPhone orientated iOS app.

Here again, a certain degree of patience is going to be of great benefit. The first time you go through this stuff, it can seem like a long, drawn out, and complicated series of steps. Moreover, it is a process that evolves through time; I think it is fair to say that most developers find small differences with every app they submit. So keep an eye on Apple's documentation, which will be a more reliable source of information than blog posts found through internet searches, which may be out of date a short time after publication.

Take it slowly and, if something unexpected happens, the web is there to help you.

# Phone Functionality

For the time being, Apple's policy on apps for the Apple Watch is that they must be additions to iOS apps and must offer some meaningful functionality on the iPhone itself. For obvious reasons this book has dealt with the WatchKit side of things and, if your apps are similarly geared to offering an experience on the watch, you'll need to think up at least a modest role for the iPhone.

My own experience with this is that it has often thrown up new ideas around phone functionality and has spurred the reappraisal of the roles of the watch and phone as parts of one gadget that just happens to be located across two physical devices.

Accept the challenge; your phone is as capable of enriching the watch app as vice versa!

# Apple's guidelines

The necessity for the iPhone app's functionality brings us rather nicely to the subject of sticking to Apple's guidelines with respect to apps that are to be submitted to the App Store.

This is Apple's game and you do need to accept that the game comes with its own rules. Some developers may view these as restrictive, others may be of the opinion that the Apple iOS and watchOS platforms have been so successful because Apple has not permitted a development free-for-all. Either way, your app needs to be accepted in order to become available on the store, so you'll need to play by the rules.

The general review guidelines that pertain to all platforms are available here:

```
https://developer.apple.com/app-store/review/guidelines/
```

With regards to the Apple Watch, almost all of this is perfectly uncontroversial and unobtrusive stuff. Things like not referring to the "Apple Watch" as "AppleWatch" or "Apple-Watch" or whatever. The documents outlining the guidelines are available here:

```
https://developer.apple.com/watchos/submit/
```

It pays to read, in advance of your submission, some of the more common reasons why app submissions get rejected, which are the subject of the following document:

```
https://developer.apple.com/app-store/review/rejections/
```

And one final thing regarding the rejection of submissions: It's great to get it right first time, if only due to the time saved by not having to do it twice and having to wait again for the submission to be assessed. The time the review process takes varies from a few days to a few weeks (although typically a week). But it's not a once-only thing, you get to try again. As many times as it takes. Apple makes its reasons accessible online, and you can fix whatever caused the rejection and resubmit.

# Keep up to date

Some aspects of the submission process seem to change every time you submit an app, others change only now and then. A small number of things haven't even changed at all. The requirements for submission to the App Store vary as new devices and new features become available, such as new screen sizes, or the addition of video previews, so you do need to stay across the latest information.

## Membership and certificates

Make sure your membership of the Apple Developer Program is up to date. If it's not, you won't be able upload anything.

Also keep an eye on your distribution certificate and be sure to review any contractual changes that need your agreement.

# Preparing for submission

Submission of an app to the App Store requires some preparation in three areas:

- The Member Center of Apple's Developer website
- Xcode's project settings
- iTunes Connect, Apple's portal for app administration

As stated already, this section is not intended to be a catch-all guide to App Store submission, but rather aims to provide the context in which additional watch-related steps must be taken.

We are assuming here that you have registered as a developer, that your Development Program membership is valid and up to date, and that you have currently valid development and distribution certificates. Please refer to Apple's documentation for further information.

# Apple Developer Member Center

The first thing we need to do is create a distribution provisioning profile for our app with the following steps:

1. Log into the Member Center at `https://developer.apple.com/membercenter/index.action` and click on the **Certificates, Identifiers & Profiles** link.
2. Select **Identifiers** from the **iOS Apps** list.
3. On the left of the following page, **Identifiers** should already be selected.
4. Click **+** to create a new **App ID**.
5. Set the new ID's **Explicit App ID** value to the bundle identifier in Xcode's project window under **General | Bundle Identifier** as illustrated here:

6. Click **Continue**.
7. Review the details you have entered and, if all is well, click on **Submit** and then **Done**.
8. Now select **Provisioning Profiles | Distribution** from the list on the left of the page.
9. Click on **+** to create a new provisioning profile.
10. Select **App Store** from the **Distribution** list and click on **Continue**.
11. Select the new App ID you just created in the steps above and click on **Continue**.

12. Select your **iOS Distribution** certificate (not the **iOS Development** certificate) and click on **Continue**.

13. Enter a name for the certificate, something like "MyApp Dist", into the **Profile Name** text field and click on **Generate**.

You can now download the certificate. Once you have double-clicked the downloaded file, Xcode will launch (if it is not running already) and add it to its list of available provisioning files.

# Xcode distribution settings

Switch back to Xcode, bring up the project window, and perform the following:

1. Select the **PROJECT**'s **Build Settings | Code Signing**, as illustrated here:

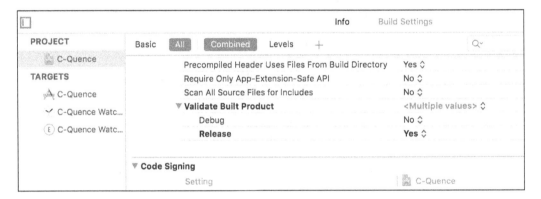

2. Select your iOS Distribution Certificate from the **Code Signing Identity** list.

3. From the **Provisioning Profile** list, select the new one that you have just downloaded.

Your app will now be code-signed with a distribution certificate. The whys and wherefores of the whole certificates thing could be the subject of a book in itself, but suffice to say here that, with this certificate in place, you will be able to upload your app to the App Store once you have setup the app in iTunes Connect.

If you don't see the new provisioning file, follow these steps:

1. Go to **Preferences | General | Accounts** and select your Apple ID.

2. Make sure you are signed in.

3. Click on **View Details** under **Provisioning Profiles** and click the new profile's **Download** button.

4. Click on **Done** and close the Preferences window.

# Using iTunes Connect

iTunes Connect is the portal through which you deal with all matters pertaining to your app on the App Store and requires membership of the Developer Program. As you can see in the following image, it offers you everything you need in terms of uploading your app in addition to the administration of all aspects of various tasks involved in dealing with the App Store:

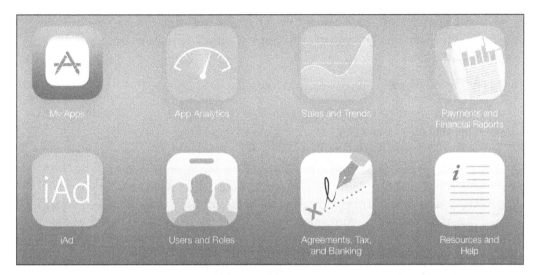

## Requirements for iTunes Connect

Before we go any further, we'll take a look at the information and resources that you'll need to complete the iTunes Connect app setup process. Preparing this stuff can be a lot of fun, you are about to become product manager, graphic artist, and publicity guru, all rolled into one.

## App description

Take some time over this, which in effect means do it now before you start the submission process. This isn't the place to discuss the do's and don'ts of advertising your app, except to say—do give it all the attention it deserves and don't leave it to a hasty ad hoc burst in the middle of the submission.

## Support URL

You will need an internet page somewhere to support your app. The level of support you offer, from a single page of how-to text, up to a fully fledged marketing and support website, is entirely up to you, but you do need the URL for submission. It is also up to you where you host that URL, so it can be, for example, a GitHub page, or a freebie web page, it doesn't need to be anything flash.

Of course, your users may form an opinion on your app's professionalism when visiting that URL (if they do), but that's a different discussion entirely.

## App Store Icon

The App Store needs a single icon image file, which will then be resized and reformatted as necessary. This doesn't actually have to be the same as the icon on your iOS (or watchOS) home screens, but it it stands to reason that this will probably be the case.

The specifications for the image are as follows:

- The size must be 1024 x1024 pixels. Remember what we said about re-using the image that you create for the watch icons? Bet you're glad you followed that advice and started with a 1024 x 1024 pixel image.

- The image resolution must be 72 dpi, the color format must be RGB, the image must be flattened (i.e. contain only one layer) and contain no transparency. If all of that sounds like you need a PhD in graphics, simply export a `.png` file of your image and 99% of the time you'll find that these are the default settings. Using OS X's own `Preview` app is a pretty safe bet.

- The image must be in PNG or high-quality JPEG format. In all your graphics work with Apple platforms, you'd be well advised to stick with **.png** files.

- The system will apply a circular mask, your image should be square (but containing no important visual elements in the corners, where they would be lost). Do not upload a circular icon.

- If your icon is designed with a white or light background, a hairline stroke will be added for display on the App Store, so there is no need to add anything of that sort.

## Screenshots

- You can upload up to five screenshots for the WatchKit app, in addition to the screenshots for the various iPhone screen sizes.

- The screenshots must be 312 x 390 pixels and should contain only the screen of the watch – do not embed it in a visual representation of the watch itself.

- The first two screenshots are immediately visible on the App Store carousel, the others become visible when scrolled to, so use those first two images wisely.

- You can choose the order that the screenshots appear in and you may wish your screenshots to tell a story. Certainly you will want to give the potential user a clear and engaging idea of what your watch app has to offer.

- Don't include any visual material for which you don't own the copyright, including logos and trademarks. This will get your app rejected.

# iTunes Connect optional data

There are also a number of optional items that you may wish to include.

## App Preview

It is possible to upload a movie file of your app in action. This preview should be between 15 and 30 seconds long and is displayed as the first image on your App Store iPhone product page, followed by the app screenshots.

Although it is not yet possible to add a video of your Apple Watch app, it may well be worth considering whether it makes sense to include at least some footage of your iOS app interacting in some way with the watch, even if that means no more than showing some UI element that clearly implies the presence of the watch app.

Making such a video requires no software that you do not have already and could hardly be easier: Just connect the iPhone to your Mac via USB, launch OS X's *Quicktime Player* app, and select **File | New Movie Recording**. In the window that opens, your Mac's built in camera will be preselected as the active input, so you'll need to select your phone using the disclosure triangle next to the **Record** button, as shown here:

You can then hit the **Record** button and begin to reveal to the world not only how great your app looks in use, but also the fact that the watch app is included.

## App Review Information

You also have the opportunity to add some comments to the Apple reviewer who is handling your app's submission. Depending on the nature of your app, you may wish to include important information here, to avoid any misunderstandings and the subsequent rejection of your submission. This may include confirmation that you are the copyright owner of any visual elements like book covers and websites: it may mean giving the reviewer access to a user account if your apps require a login, or something equally essential to fully reviewing the app.

 A few words of clarification could save you a delay of weeks in releasing your app, so do make use of this opportunity.

# The iTunesConnect process

We will now prepare iTunes Connect to expect the upload of your app.

1.  Go to `https://itunesconnect.apple.com` and sign in, using your Apple ID.
2.  Click on the **My Apps** link.
3.  Click on **+** and select **New App**, which will allow you to enter the resources and information necessary to enable your new upload.
4.  Fill in the **Name** text field; this will be the name as it will appear on the App Store. Your app name must be unique - you can't use the same name as another app on the Store.
5.  From the **Primary Language** drop-down menu, choose your app's primary language.
6.  From the **Bundle ID** drop-down menu, choose the new provisioning profile that you created in the Member Center (and which you selected in Xcode's Build Settings).
7.  The **SKU** is an ID for your administrative use only, nobody else sees it. It must be unique among your apps and can include letters, numbers, hyphens, periods, and underscores. Use it for whatever purpose you choose, but if you don't use it, you still need to fill in this field.
8.  Click on **Create**.

You will now be taken to the **App Information** section of App Store Information Page. The **Bundle ID** should be correctly selected, so you just need to choose your **Category** (and optional **Second Category**, if you wish) and then click on **Save**.

Now move on to the **Pricing and Availability** section of the App Store Information Page and name your price.

Select the **Prepare for Submission** section of the App Store Information Page.

You should already have everything you need for this section, as outlined in this chapter.

## Submit for review

Ready? See that button, **Submit for Review**?

Savor this moment. After you've hit that button, your app's on its way. All you have left to do is use Xcode to upload the binary (the app file that you build) and you're done.

Go ahead. Submit!

# Uploading the build with Xcode

Once you've added the app to iTunes Connect, it'll show the status **Prepare for Upload** on the app summary page. Click on **View Details** then, on the app info page, click **Ready to Upload Binary**.

Answer the question as to whether your app contains encryption and click on **Save**. Now, on the app summary page, your app status should be **Waiting for Upload**.

# Upload

In Xcode, select **Product | Archive**, as shown below:

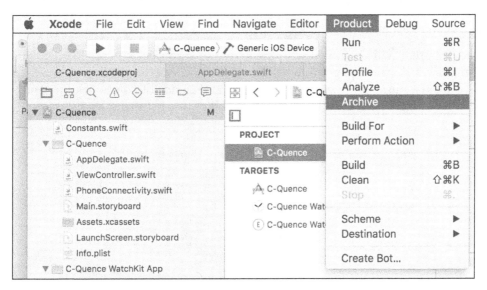

If the **Archive** item is grayed out, you probably have a Simulator device selected under the **Run** settings, so select **Generic iOS Device** instead (see illustration above).

The archiving will last a few moments and then you'll be presented with the Organizer window, with which you can upload the archive to the App Store.

# Uploaded!

Congratulations! You're finished.

All there is to do now is wait for the outcome of the submission. If you have adhered to Apple's guidelines, you should soon get a mail informing you that your app is ready for sale (even if it's for free).

# Summary

That was quite a journey. If it's any consolation, the process has become shorter, simpler, and less error prone as Xcode has developed as an Integrated Development Environment, and iTunes Connect and the Member Center have been tweaked and made more intuitive.

But that was the final step to getting your app released and the next chapter will deal with some of the things you'll need to consider post-release, as well as looking at the many ways there are to go from the point you have now reached.

You have now successfully planned, developed, and tested an Apple Watch app and submitted it to Apple's App Store.

Congratulations again for making it this far!

The next chapter will equip you with the tools and resources you are likely to need moving forward from this point, whether you will work alone or as part of a team; it will look at a number of ways to improve the efficiency of your workflow and offers a few suggestions for adding that extra bit of magic to your app.

# 10
# This Is Only the Beginning

That was nine chapters of pretty intense material. You will probably (and quite rightly) be proud of your achievements so far. From now on, everything you learn will be built upon a solid foundation that has already enabled you to get an app submitted to Apple's App Store.

This chapter is concerned with the many possible avenues that are available to you to deepen and broaden your programming skills, to make the most of the valuable time you spend coding, and to give your app the best possible chance of achieving its potential once it has been released to the world at large.

In this final chapter we will cover the following topics:

- Advanced custom navigation
- Onboarding
- Using the Code Snippets Library
- Post release improvement and maintenance
- Analytics frameworks
- Program design and programming paradigms
- Using the command line
- Using Xcode's tools
- Git and version control
- Useful utility apps and essential resources
- Open sourcing

# Using animation to the fullest

We have seen how animation can be used to produce a more vibrant, engaging user interface, but this is only scratching the surface of what can be done by animating the changes in the property values of UI elements. Despite the relatively modest number of properties we can animate, with a little imagination we can create effects that are simple, but which fundamentally alter the way we engage with the Apple Watch.

We have already seen in *Chapter 2*, *Hello Watch*, how simple it is to overcome the lack of a completion handler in calls to WKInterfaceController class's animateWithDuration method by extending the class with a method that chains together sequentially an arbitrary number of animation blocks. By combining concurrent and sequential animations we have at our disposal an unlimited number of ways to add complex (although often subtle) movement to our interface without having to deal with masses of complex code. The key here is to think about what we can do with the techniques at our disposal, rather than to search for new techniques or third party libraries (or to not try in the first place).

# Advanced custom navigation

Of course, animations don't have to be subtle. If we consider that the watch screen is fundamentally an object that shares many characteristics with WKInterfaceGroup itself, we begin to approach the idea that, conversely, a WKInterfaceGroup object can be considered to be a container for a whole screen of user interface elements. Such a screen is not subject to the limitations of the top level WKInterfaceControllerview, and by swapping between several such UI groups, we can effectively extend our navigation palette far beyond that which is offered by vanilla WatchKit.

Using the techniques that we have covered in many of the previous chapters, you are able to add two or more WKInterfaceGroup objects to the InterfaceController in Interface Builder. Put them both in an outer WKInterfaceGroup, making sure to set its **Layout** property to **Horizontal**. This group will contain everything in the UI, as illustrated here:

By setting each of these groups, let's call them UI-Groups, to be the size of the whole screen, that is, by setting both their **Height** and **Width** properties to **Relative To Container** (with the default vale, which is 1.0), we can change screens using any of the size and layout properties of those objects.

In the GitHub repository, there is a small Xcode project called `AdvancedNavigation`, which will illustrate a very simple implementation of this idea:

`https://github.com/codingTheHole/BuildingAppleWatchProjectsBook`

All this project does (as you will see if you download it) is change the widths of two side-by-side UI-Groups. The one that should be visible has a `RelativeWidth` value of 1.0 and the other, rather than being hidden through its `Hidden` property, has a `RelativeWidth` value of 0.0. The values of these properties are then swapped to effect a change in which UI-Group is visible; by wrapping this in an animation closure, we create an interesting transition from one screen to another. In the example code in the GitHub repository, each UI-Group contains nothing more than a button to start (or revert) the change of screen along with an image view to make things slightly clearer to follow, but in your app such a UI-Group could contain further buttons, labels, tables, even the `WKInterfaceMap` objects, as well as further `WKInterfaceGroup` objects. There is no limit to the flexibility of an interface constructed in this manner.

# Onboarding

One popular trend in UI design, and one that is to be widely applauded in this author's opinion, is to include a so-called onboarding experience in an app, by which the user is guided through the essentials of an app's UI, usually through a sequence of static screens. Once the guided tour is over, the user is familiar with a number of features that the app has to offer, as well as a good idea of how to use them. The onboarding sequence is never seen again.

This is an excellent way to introduce your users to your app.

You might think, hey, that's a lot of resources to include in my tiny watchOS bundle for a feature that will only be used once, but in fact the code that defines a storyboard is no more than a modest amount of text in XML format. For example, here is about half of the storyboard from the project from *Chapter 2*, *Hello Watch*:

```
<?xml version="1.0" encoding="UTF-8" standalone="no"?>
<document type="com.apple.InterfaceBuilder.WatchKit.Storyboard"
version="3.0" toolsVersion="8173.3" systemVersion="14F27"
targetRuntime="watchKit" propertyAccessControl="none"
useAutolayout="YES" useTraitCollections="YES"
initialViewController="AgC-eL-Hgc">
<dependencies>
<plugIn identifier="com.apple.InterfaceBuilder.IBCocoaTouchPlugin"
version="8142"/>
<plugIn identifier="com.apple.InterfaceBuilder.IBWatchKitPlugin"
version="8089"/>
</dependencies>
<scenes>
<!--Interface Controller-->
<scene sceneID="aou-V4-dly">
<objects>
... more code here ...
</document>
```

Not much, is it? So creating a series of Group objects is an inexpensive way to add value to your app (provided that you don't pack them full of images and movies, of course) and represents a perfectly viable way of creating a pleasing, helpful, and ultimately rewarding introduction to the user's experience with it.

Once the sequence has been viewed by the user, simply add a `Bool` value to the `NSUserDefaults.standardUserDefaults()` dictionary, indicating that this is the case. Each time the app launches, it checks whether this flag has been set to `true`, and if it has, it skips the intro.

These are just two of the ways that you can leverage WatchKit's simple but powerful animation features to create features that would normally be associated with the larger devices without using up a ton of memory and writing horrifically complex code.

# Making use of code snippets

Lots of code that you write gets written over and over again. A very convenient place to store frequently used code is in the Code Snippet Library (*Command-Option-Control-2*).The image below shows the details of a snippet that has been assigned a keyboard shortcut. When this snippet is inserted, the cursor is placed at the ... text, ready for typing.

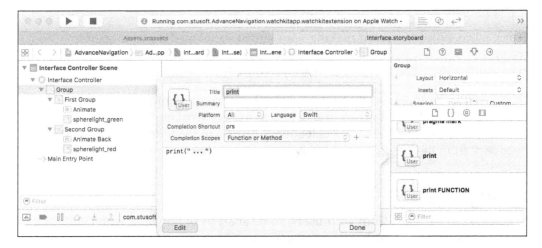

To create a new snippet, select the relevant code in the source code editor and then drag it onto the Code Snippet Library.

After a while, you'll find that you have a substantial number of keyboard shortcuts at your command for typing code that you use frequently. Although the example shown above is very short (and trivial), there is no reason not to create much larger snippets, which, for example, could stub a set of the `WKInterfaceTable` callback methods.

# Post release maintenance

Your app may be in the Apple's App Store, but there is still plenty you can do to improve your users' experience, as well as the app itself.

## Support

Good after sales support is as important as good code to the success of your app, never more so than in the period immediately following its initial release. Try to make as much use as possible of the resources at your disposal, including your support website and social media. The exact way in which you do this will vary hugely according to the nature of your app and the type of user you expect to be engaging with it.

Think twice, however, about making your site or social media pages open to comments from users. One disgruntled user can do a huge amount of damage, even while a thousand contented customers quietly use your app every day.

Direct feedback can, however, prove extremely useful in identifying the areas of your app that could use improvement, or that are not working as you intended them to. It is worth considering adding a feedback feature to the iPhone companion app (feedback from the watch itself is likely to find few takers, one would imagine).

## Analytics

In addition to feedback from your users, some analysis of how and when your app is being used could be critical at this phase. iTunes Connect now offers a considerable degree of analytical data about your app and you should most certainly take advantage of this free and constantly evolving source of information.

Two other popular candidates are also worthy of mention:

## Google Analytics

GA provides a free iOS framework that can be tailored to your app's needs and woven into its very fabric. You can set up GA to report particular events, window closures, engagement times and a thousand other things. The trick here is to decide what you actually need, and will profit from expending time on, both in including the code in your app (which is very simple but can become quite pervasive) and in reading through the data.

The documentation and download links can be found here:

`https://developers.google.com/analytics/devguides/collection/ios/v3/`

## Fabric/Crashlytics

Another free framework, this one dedicated to crash reporting. If ten thousand users are using your app, you will be able to gain a very accurate picture of where any crashes are occurring, which will be a valuable aid in finding and fixing them.

If you choose to include this tool, here is where to start:

`https://get.fabric.io`

Basically, for post release, all these strategies are leveraging the fact that your app is now in active use, which can give you insights into its life among users that cannot be gained in any other way. It will cost a little time to include the code, and considerably more time to interpret the data, but some degree of analytics is probably appropriate to almost all mobile apps.

# Expanding your skills

One of the great things about programming is that we will forever be able to expand our skills. New platforms (like watchOS) arrive, new programming languages are released while the old ones evolve, programming patterns change, and some practices inevitably lose their relevance.

Software engineering, like engineering in general, is too broad a field to master completely. Gone are the days when a skilled programmer would know most of what all the other skilled programmers know, and to a certain extent, we must all choose what we will learn and what we will not. It is probably wise to keep up to date on a core set of topics and technologies that pertain to your interests and activities as a programmer, however, given our common position as WatchKit developers, I'd like to suggest a couple of pointers.

# HTTP

Being mobile means talking to the internet and talking to the internet means mastering at least the basics of the hypertext transfer protocol, better known as HTTP. While we have touched on this topic in this book, it is worth spending some time expanding your knowledge of not only HTTP requests et al, but also of the `NSURLSession` framework that provides so much of the functionality that your mobile apps will use frequently.

The more comfortable you are with this topic, the better equipped you will be to meet the demands and challenges of an increasingly connected environment.

# Swift

As has been mentioned elsewhere in this book, the Swift language is one that is still evolving. At the time of writing, Swift 2.2 is nearing release and a roadmap for Swift 3.0 is being discussed by Apple and the community at large, now that Swift as been made open source.

Apple's own publication, *The Swift Programming Language*, is available for free in iTunes and can be read using the iBooks app. It is part tutorial and part reference work, and is an excellent starting place to learn Swift. It is kept up to date and iBooks will download new versions as they are published.

But you probably know that already.

# Program design topics

When developing apps, the programming language you use is far from everything. Most languages offer you considerable flexibility in terms of what kind of patterns you use to create programs from those languages, and the ways in which small areas of functionality are woven together to form fully fledged applications.

# Programming paradigms

It is only natural to assume that the way we have learned to do something is in some sense the 'natural' or 'obvious' way to do it and software development is no exception. But beyond the style and practices followed in this book (and the majority of Swift orientated blog posts and code samples), there is a whole universe of doing things differently, sometimes completely differently.

Prepare to be amazed.

## Object orientated programming

Let us first look at what we have covered in these pages in the context of the broader programming picture.

In the mobile space, Objective C, Java, and of course Swift, are three of the most popular programming languages, and all three of these languages share a large number of characteristics with respect to assumptions that are made about how the software they are used for writing will be structured. In all three of these languages (and many more besides), we tend to combine a number of functions and variables into well defined groups of code that we refer to as objects, of which we may create as many instances as we require. These have traditionally been expressed as classes, encapsulated areas of code that hide their internal workings from the rest of the code, exposing only a limited number of methods and properties though their Application Programming Interface, or more commonly, **API**.

This is the fundamental basis of program construction; objects are created, which may contain or create other objects. They may be handed around to other objects to be manipulated or read and are frequently sub-classed into objects that represent further specializations of the superclass's code. This style of programming is called Object Oriented Programming(OOP).

Now, Swift is a very modern and flexible language, and we will see shortly that it is capable of more than just OOP, but it is nevertheless the case that most code written in Swift, including the code written in this book, is object orientated, which would explain any reaction the reader may have along the lines of, well of course we use objects and subclasses and all. What else is there?

We will look at two general groups of programming paradigms (i.e. fundamental programming styles), **imperative** and **declarative**. Imperative programming includes object orientated programming, so we'll start with that.

## Imperative programming

Imperative programming means basically telling a program what you want it to do by writing a list of statements, that is, *do this*, then do *that*. Let's take a couple of lines of code as an example:

```
let a = 0
let b = a + 1
```

What we are writing here is basically telling the compiler:

1. Create something called a and assign to it a value of 1.
2. Evaluate the expression a + 1.
3. Create something called b and assign to it the value returned the expression in point 2.

Step by step, we are telling the program exactly what to do, thus the name *imperative*. We will contrast this shortly with declarative programming. But OOP is not the only imperative style around. It's not even the only imperative style available to development in Swift.

Other imperative paradigms include the following:

**Protocol Orientated Programming** is the new buzz-word in the Apple community. After an entertaining and very clearly explained presentation at the WWDC 2015 by David Abrahams, the iOS Swift community has been eagerly exploring the possibilities and advantages of using protocols to share behavior between objects instead of subclassing.

 The WWDC video sessions URL is listed below in the section *Watch This*.

It is really worth watching, if for nothing else than discovering Crusty and his contribution to software engineering.

**Machine Code and Assembly Language** refer to languages that go right down to the metal and deal directly with the 1's and 0's that computers manipulate. In the case of assembly languages, there is a thin layer of abstraction, but, basically, this is programming in its rawest form. It's fun to try out, and it's essential to many embedded systems that rely on direct control of hardware, but you seriously don't want to be writing a web browser in machine code.

Procedural languages include C and Basic, and are basically imperative languages that don't support the use of objects to encapsulated code, so no subclassing and the like. They were the predecessor to object orientated languages such as C++, and are still in use today, particularly C. If you ever get involved in server side programming, for example, you are likely to need to be comfortable with C.

You would be right to think that the programming you have done in Swift includes vast quantities of such code, but the point is that these languages stop there, they offer no support for anything more. Thus, they are termed procedural languages.

## Declarative

So what is the alternative to telling a computer what to do, step by step? Hinting politely at what you want it to do?

Well, there is a grain of truth in that. Declarative programming means that you are telling a program *what* you want it to do, rather than *how* you want it done.

Consider the following: You can put a driver in a car and then give him turn-by-turn instructions on how to get to wherever you want to go. Every corner, every traffic light, perhaps even which foot to use to brake and accelerate. It will be a long list of instructions, to be sure, but you will get there in the end.

Alternatively, you can give the driver a map, name the destination, and sit back and enjoy the journey (or the conversation—I come from a city of very loquacious taxi drivers). Declarative programming involves describing the goals rather than the steps taken (though the *how-to* sections of code will possibly include short bursts of imperative subroutines). Code written in this style is often much easier to read, since it is possible to follow what a program is doing without wading through the intricacies of the steps it takes to do it.

SQL is a popular declarative language with which you may be familiar.

**Functional programming languages** like **Haskell** use this idea and combine it with the idea that a function should give the same output for a given input every time, which means removing any ability to refer to properties of objects or as we should call it, **state**. Variables are assigned only once (and are then not very variable, but there you go) and are said to be immutable, which sounds like an odd concept coming from an OOP background.

We can't do justice here to this fascinating topic, which has contributed so much, incidentally, to the development of Swift, and the reader is strongly encouraged to have a good look at what functional programming techniques can offer the modern developer, particularly the developer who has the good fortune to be working in Swift. It is also true to say that two or three days of learning a pure functional programming language such as Haskell will unlock much of the potential of many of Swift's features, and provide a mental framework for ideas that are a valuable supplement to object orientated programming techniques.

Logic programming, as supported by, for example, Prolog and symbolic programming as with LISP are also ideas that are worth at least some brief exploration, though probably less so than those listed above.

## Where Swift fits in

Swift is a multi-paradigm language, as are others such as Python and JavaScript. Object orientated, functional, protocol orientated, mix and match them in Swift as you see fit. It is still a very new language, it is impossible to predict how programming practices will develop over the next decade, and this is a really exciting time to be part of that development.

Certainly, you could remain ensconced within the object orientated paradigm and still write first class code, and it may be the case that you're more inclined at the moment to get more of that under your belt before venturing into other paradigms and even other languages. But in the end, the borders we draw between these patterns and paradigms are artificial ones, and there is absolutely no reason why a little functional knowledge shouldn't be as easy to gain and as valuable to have as a WatchKit framework that is new to you, some third party code that will save you reinventing the wheel or any other concept that may seem 'closer to home'.

# Program design patterns

We have had a brief introduction to the **Model View Controller** (**MVC**) program design pattern espoused by Apple, but it's not the only one by far, whether inside the object orientated paradigm or outside it. While these patterns may or may not spark your interest, it is worth having at least a passing familiarity with their broad concepts, and an alternative view of how things could be done is always of value anyway.

A two-sentence introduction here would be of very little value, and an in-depth look at these patterns is unfortunately beyond the scope of this book, so the reader is encouraged to research these topics online. Here are a few search terms to get you started:

- MVVM (Model-View-View-Model)
- MVP (Model-View-Presenter)
- Decorator pattern
- Strategy pattern
- Singleton pattern
- Factory method pattern

As with the programming paradigms, these are not topics that you need to master completely and immediately, but some familiarity with them will put your own current coding habits in perspective and will give you a much better view of the bigger picture as you expand your skills as a developer.

It is probably clear from this section that you can't do it all, nobody can. Try to maintain a balance between a broad general knowledge of what's going on in the development world and a deeper, more localized area of expertise in which you are likely to quickly attain a level of knowledge beyond that possessed by most programmers.

Your own interests, strengths, and intuitive inclination will guide you in the path you take and rest assured that however obscure your tastes and skills may be, somebody somewhere needs them.

# Tools

Typing code into Xcode and hitting the Run button to see if it works is only a part of the story (though undoubtedly the most central part). There are a number of other tools that you should consider an essential part of your toolbox and countless others that may be of immeasurable help to you, once you have found them (often easier said than done) and learned what they can do for you.

What follows is a brief look at some of the most important tools already at your disposal (in that they are already installed on your machine), some further utilities that are free to use, and one or two paid apps that are at least worth checking out and are not expensive to license should you find them useful.

# Terminal

In case you have not met it before, Terminal is a Unix console (Unix is the basis of OS X) and is installed along with the operating system. It is an important tool in many ways, though its value really first becomes apparent after you have become familiar with using it.

The Unix commands that can be run through the terminal are something that you'll need to get used to sooner or later and beginning early is the key to a successful integration into your workflow.

 The terminal interface is also referred to as the console or as the command line.

# Help from afar

One important use case for the terminal is that, very often, help offered from developer forums will need at least a modicum of ability on your part to use the command line. This might mean no more than deleting a rogue file or checking which version of some command line tool you have, but you don't want to be bugging people about trying to do it in the Finder. You're a developer, talking development with other developers, and there are certain expectations around your ability to deal with your system via the terminal.

# Command line development

Very often it is more convenient to check out some parts of your code from the command line instead of building the app and running it followed by a load of navigation to the screen that contains, say, the button that calls the function you want to test. Calls to resources on the web, typically an HTTP request for an asset located on a remote server, are excellent candidates for this use of the terminal.

 In the command line input that follows, the tilde, ~, represents the location of the folder that Terminal is 'looking at'. It is the prompt that shows that Terminal is awaiting instructions, and you do not type it as part of those instructions. In other publications and posts you might see this prompt expressed with the $ character, or >>>.

## Using cURL

Firstly, let's just check the content of an online text file. There are a number of situations in which we may wish to inspect a file's contents before we write the code that will handle the HTTP response and, although we could write code into our app that does that, this way will be much quicker.

The terminal already knows how to do this, using the cURL library, so you don't need to write any code so much as call it. Type the following into the terminal (omitting the tilde, remember):

`http://www.grimshaw.de/applewatchprogramming/sampleContent.txt`

The response (which is the contents of the `sampleContent.txt` file at the location `http://www.grimshaw.de/applewatchprogramming/`) will be printed to the terminal:

`Congratulations, you've successfully used curl to access content on the web.`

How easy was that? The next time you need to see the contents of a JSON resource in order to know how to handle it in your app, you'll know how to access it simply and quickly.

## Creating a local server

Sometimes, it's more convenient to be able to serve up JSON, HTML, and other app assets, especially huge media ones like movies, from your own hard disk, rather than downloading them from a remote server, particularly for testing purposes. It's faster than downloading, it doesn't use up traffic, and it's much easier to add, change, and delete files using the finder, than pleading with a client's system administrator to delete the file you just ask him to add to a directory.

So any resources you can locate on your hard drive are going to be a great help. But you'll need to turn your Mac into an HTTP server!

There are several ways to do this, some easier than others, some really easy. Using the Python language, it's as easy as one command.

> Do I have to learn Python now? I hear the reader cry.
>
> No. The single line of code is all that we will need to get our local server up and running and logging the traffic that uses it.
>
> However, given the importance of mastering HTTP, you might wish to visit the following Python documentation page, to learn what else you can do with the server once you have started it:
>
> https://docs.python.org/2/library/
> simplehttpserver.html

Type the following on the command line:

```
~ python -m SimpleHTTPServer 8000
```

This creates a server at the ~ location (home) on your hard drive, using port 8000 (stick with this port unless you know you have a reason not to)

This time you'll see a response in the terminal window that looks like this:

```
Serving HTTP on 0.0.0.0 port 8000 ...
```

Now, if you put any file, say a .txt file, in your home directory (because that's where you started Python's server, remember the tilde), you can access it from the command line with the following:

```
curlhttp://localhost:8000/yourFile.txt
```

You can also access it through *Safari* or any other web browser using this URL in terminal:

```
http://localhost:8000/yourFile.txt
```

in Terminal

> You can start the server from any directory (that is, folder) on your drive that you choose. The simplest way to open Terminal pointed at a particular directory is to drag that directory onto the Terminal application icon (in the Dock is easiest).
>
> However, you should get used to using the cd path/to/your/directory command in the console as well. cd is a Unix command and stands for Change Directory.

You would then access a file in that directory with the command:

```
curl http://localhost:8000/path/to/your/directory/yourFile.txt
```

The Terminal window in which you launched the server will show you a log of the HTTP activity that has taken place, as illustrated below:

```
● ○ ○              ⌂ stu — python -m SimpleHTTPServer 8000
[~ python —m SimpleHTTPServer 8000
Serving HTTP on 0.0.0.0 port 8000 ...
127.0.0.1 - - [21/Dec/2015 10:56:23] "GET / HTTP/1.1" 200 —
127.0.0.1 - - [21/Dec/2015 10:56:45] "GET / HTTP/1.1" 200 —
127.0.0.1 - - [22/Dec/2015 12:55:38] "GET /hi.txt HTTP/1.1" 200 —
```

# Xcode's Instruments app

Instruments, which is included as part of the Xcode installation, comes with a number of tools that are there to help make developing apps easier, and it is a good idea to start using them, in however rudimentary a fashion, as soon as you can.

Information from these analysis tools includes:

- Graphics operations and performance
- Object and other memory-related allocation statistics
- Memory leaks
- Information about File system reads, writes, and other operations
- Statistical samples of your application at runtime
- Process-specific and system-level activity

This is by no means an exhaustive list. The important thing here, however, is to make a start with using these tools. You don't need to become an expert to make use of the information that they provide, and the sooner you start scratching the surface, the better.

Apple's own example of how to use the Instruments application is here:

```
https://developer.apple.com/library/ios/documentation/Performance/
Conceptual/PerformanceOverview/InitialEvaluation/InitialEvaluation.
html#//apple_ref/doc/uid/TP40001410-CH206-SW7
```

The full User Guide is here:

```
https://developer.apple.com/library/ios/documentation/DeveloperTools/
Conceptual/InstrumentsUserGuide/
```

# Application Loader

```
https://itunesconnect.apple.com/docs/UsingApplicationLoader.pdf
```

This app is available for download to registered developers, and is an alternative way to upload binaries to the App Store, bypassing Xcode. I don't use it often, but it's useful to be familiar with its use, especially if you ever find yourself in a situation in which you are compiling the builds, but someone else is doing the uploading to the Store using Application Loader, which happens in a lot of medium sized and large companies.

# Version control

Next to Terminal, this is another technology independent of Xcode that you'll want to be on good terms with. If you ever write code for anyone but yourself, this will be an important skill. If you only write code for yourself, this is an equally important skill.

## What is version control?

Version control systems keep a record of changes you make to your project. You **commit** the code to a **repository** at regular intervals. This is the equivalent of taking snapshots with Xcode, but it can do a lot more. Several people can access a repository, or repo, and version control will keep tabs on who did what and when. You will have a complete record of all the commits ever made, by everyone on the project.

You can also revert to an older version of your project if you ever need to refer back to code that was deleted or overwritten at any point.

# Git and repos

There are different version control frameworks out there, but there is one that is far and away the most popular and that is called Git. Developed by Linus Torvalds, the guy that wrote Linux, Git is simple to use at beginner level, but offers all the advanced features you could wish for once you get to be a power user.

You can create different branches of the repo to develop code away from the main branch, which remains then in a stable state during feature development. Any time you wish, you can merge a branch into any other, thus updating, say, the Master branch, once development on the *FeatureX* branch (perhaps worked on by a different developer) is considered complete and safe enough to add to the release candidate version of the app.

Thus Git is a backup solution, collaboration tool, and history log all rolled into one free package. The only free software bigger and better than that is probably Linux itself.

# Git and Xcode

Git version control is built right into Xcode, so you have a combination of Xcode, the command line, and access through a web browser (see *GitHub* and *BitBucket* below) to create, update, and administer all the commits (by all of a project's developers, should there be several) of all the projects you have in all of your repos.

To commit your code (that is, take a snapshot), type *Command-Option-C* in Xcode, and enter the (obligatory) commit message, which you'll be glad of six months from now, I promise.

There is no room here for a fuller explanation of what Git and Xcode's version control can do for you, but do find time to learn this stuff, it's worth every minute spent on it.

# GitHub

`https://github.com`

A GitHub free account provides you with as many **remote** repositories as you want. What that means is that you have a permanent backup of your local disk's project files online, including a complete history of the commits made, and one that is available to others for download and possible collaboration (should you choose).

Just use Xcode or the command line to push the committed code to the remote repo, and you'll never again have to wonder why *featureX* of your app was working so well last month, and has at some point become dysfunctional, because you'll be able to refer to that older version of your project.

GitHub has an accompanying app named *Desktop* that you can use for administration of your local and remote repos, if you prefer that to the command line.

## BitBucket

```
https://bitbucket.org
```

Everything said about GitHub applies equally to BitBucket. It offers free accounts to host your repositories, but provides a more sophisticated OS X app (named *SourceTree*, which is a much better name than *Desktop*, if you ask me). You may find it a little too sophisticated, and my advice would be to start off using Xcode and the command line until you have gained some experience with version control.

If branching is something you do a lot, you may find that the BitBucket site's representation of what has been committed on which branches makes it preferable to GitHub.

## Which one to choose

GitHub is more popular and is familiar to almost all developers. That is the reason why the code for this book is on GitHub. BitBucket is integrated with a ton of other tools from the company that owns it, *Altlassian*, including *SourceTree*, *Jira*, *HipChat*, and *Confluence*, tools that have gained a lot of traction in the corporate world, and many developers love *SourceTree*.

Since both platforms are free, why not try both?

Whichever you choose, try to take it easy at first; get used to making regular commits, composing succinct but informative commit messages, and pushing to the remotes every time you power down or walk away from your development machine.

# Personal favorites

The next few tools are ones that I have found to be regularly helpful and that are either free or very low cost, some of which we have met before.

## Graphics

Like it or not, as a developer you're always going to need to rustle up a graphic image or two for every project that goes beyond the very early stages of development and quite possibly, all the other graphics for an app too.

I have found the following utilities to be quick to learn, intuitive to use, and provide everything I need in terms of features:

## Graphic

We met this when designing our icons in *Chapter 8, Images, Animation, and Sound.* A fast and easy way to get PNG files designed, sized, and exported, Graphic costs a few dollars, but you want people to pay for your apps too, right?

Graphic is available here:

```
https://itunes.apple.com/us/app/graphic/id404705039?mt=12
```

## iConeer

This app has also been mentioned before. All the icon sizes you need for all Apple's platforms with a couple of mouse clicks and a drag-and-drop, as shown here:

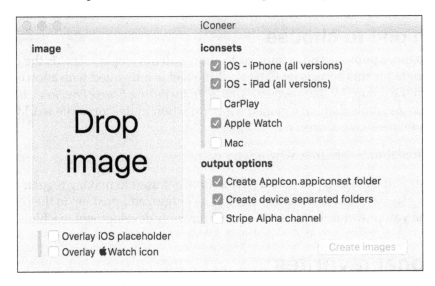

Iconeer is available here:

```
http://www.quickgets.com/iconeer/
```

## Bezel

Sometimes you need to see how your app looks on a watch. If yours hasn't arrived yet, or if you need to show your app to someone else, or if you haven't decided whether you even want an Apple Watch yet, try Bezel, which will provide an alternative to the Simulator window illustrated here:

The Bezel window, given below, puts the whole thing into context:

You can even change the color of the strap and the Apple Watch model.

Bezel is available here:

```
http://infinitapps.com/bezel/
```

# HTTP traffic

Monitoring what is going in and out of your app in terms of HTTP traffic is sometimes essential. Although we have seen that our Python server logs all its traffic in a Terminal window, we can't use that for traffic between an internet server and our development machine or a test device. To do this, we must reach for a third party tool.

## Charles

Charles provides an excellent overview of the requests being sent from your iOS device or from OS X itself, as well as the responses that are returned from the web. It is an excellent tool, but is not free, though there is a free trial period.

During development, I have Charles running all of the time, which says something about its usefulness.

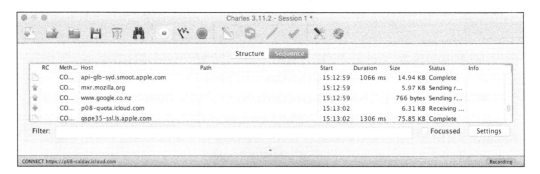

## Wireshark

Wireshark does something similar and is free, but I can't vouch personally for its features or ease of use, not having used it. It is widely recommended on the forums, which would imply that it's good at what it does.

# Sites to be aware of

In this section I am casting objectivity to the wind to list the dozen sites that I visit most frequently as a developer. It is likely that at least half of them would be in any developer's Top Ten, but that's not the point here. This is just to introduce you to a small number of sites that I personally find useful, or entertaining, or both I suppose. Some of them have been mentioned earlier in the book.

# Swift

Swift must be one of the fastest evolving programming languages out there and it pays to keep abreast of the latest developments and additions to the language.

# Open source Swift

This is the site that was set up at the same time as Swift was open sourced and is as fascinating as it is essential. There is a lot of stuff here, much of which is perfectly readable for less experienced developers, as well as material for the most advanced of software engineers:

```
https://swift.org
```

# Swift blog

The blog site run by Apple has been around since the language was released and, although it is updated infrequently, it is still a premium source of information about the Swift language:

```
https://developer.apple.com/swift/blog/
```

# Apple developer resources

It will come as no surprise to you some of my main ports of call are Apple's documentation and further sources of information.

## watchOS developer library

This is the most valuable source of information available. Today there are 972 documents available here; well sorted, well ordered, and searchable, they will answer a large proportion of questions you have around watchOS development. Whatever your level, you need this:

```
https://developer.apple.com/library/watchos
```

## WWDC vids

The annual World Wide Developer Conference in San Francisco is the event of the year for all developers working on Apple platforms. For weeks before and weeks after, nothing is discussed more on the forums, newsgroups, and blogs than the contents of the Keynote speech and the smaller, more specialized sessions that deal with particular topics around what's new in Mac OS, iOS, watchOS, and now tvOS.

Each year the sessions are filmed and made available online. The videos are a mixed bunch, depending on the topics and on the presenters themselves. Some are funny, some are truly nerdy, and some shine with brilliance (a few of them tick all three of these boxes), but what these videos have in common is the fact that they are all packed with relevant information, there is very little marketing baggage, and they're extremely well produced. Most are between half an hour and an hour long, which makes for a very rewarding burst of coding knowledge.

The videos can be found here:

```
https://developer.apple.com/videos/
```

One word of caution regarding data: if you have a fast data connection, Apple will quite happily stream gigabyte upon gigabyte of data to you, giving you insanely high definition video that you might not appreciate. So if you don't have unlimited data, be sure to download the lower resolution SD version instead or you might find yourself on dial-up speed for the rest of the month.

# Online help

It's nice to know that you're not the only one that hasn't known from birth how to change the alpha value of the border color of the `UICollectionViewCell`. The following few sites are those that I visit most often of all the sites listed here and have provided more answers over the years than everything else put together.

Don't worry if you don't understand 90% of the questions. Everyone has their niche, you'll have yours, and there's room in the world, and web forums, for a lot of niches.

## Stack Overflow

Already mentioned in *Chapter 9, Wear It, Test It, Tweak It, Ship It*, Stack Overflow aims to provide what developers crave most: clear, specific answers to clear, specific questions:

```
http://stackoverflow.com
```

And try to contribute! It's enormously satisfying to watch those Stack Overflow reputation points trickle in from people who have found your input helpful. Whatever your level of experience, you are guaranteed to know something that will help someone else with a problem they are having.

## Stack Exchange

Stack Exchange, on the other hand, provides the programmers sub-domain, which fills in the gaps around more general programming topics. Program design patterns, where to put your constants, whether singletons are evil or indispensable, it's all here:

```
http://programmers.stackexchange.com
```

## Apple forums

Another essential resource. If nobody here knows it, then nobody knows it anywhere:

```
https://forums.developer.apple.com/welcome
```

# My sites

There are two online resources that are dedicated to supporting this book:

## Support

```
http://www.grimshaw.de/applewatchprogramming/
```

The support site hosts anything and everything that might be relevant to the book but won't fit in. This site will grow with time, so it's worth checking back now and then.

## GitHub

```
https://github.com/codingTheHole/BuildingAppleWatchProjectsBook
```

This site contains the source code for the examples in the book, some supporting code that didn't make it, and sundry video, graphic, and audio assets that might be helpful.

# Best of the blogs

Whether you're an expert or a complete beginner, pro or hobbyist, someone somewhere is writing for you:

## NSHipster

`http://nshipster.com`

Originally written by the brilliant Mattt Thompson (yes, three T's) and continued by the equally brilliant Nate Cook, this blog contains some of the clearest, most enlightening, and most relevant software development writing available anywhere on the web. The posts are dedicated mostly to iOS and so highly relevant to any Apple Watch developer.

## Erica Sadun

`http://ericasadun.com`

Erica's is another wonderfully informed and erudite blog, she really does know her stuff. Hers was the first change that was accepted for inclusion in Swift 3.0 that came from the community, as opposed to Apple itself. Every post is a gem.

## Natasha the robot

`http://natashatherobot.com`

And just to prove that you don't need a decade in the business to write about it, Natasha's blog deals with her experiences in learning to program in Swift and intermediate programming topics in general. Her contribution to the community is based not on top-rank knowledge, but on an ability to convey the excitement but also the apprehension of trying to become a better coder.

# Stay in touch

Expanding your skills means keeping up with what's going on in the software development world, both in relation to iOS/watchOS coding and to broader development topics. It is essential to take time away from coding in order to prevent your knowledge from becoming out of date, but the spike in motivation you get from learning what's going on may prove just as valuable.

# Follow the buzz

Read the blogs, subscribe to the newsletters. Not all of them perhaps, but those that seem most relevant or interesting (a list that will evolve with time, of course). Many of the ideas that will shape your growth as a developer are out there, just waiting to be discovered, but you won't find them by exclusively writing code; you need to proactively seek out new topics of value to you.

Use the many internet forums out there not just for problem solving, but also as a way to keep up with issues that are affecting developers of all platforms, in all locations. Many concepts of which you are not yet aware will trickle into your mental model of the programming landscape and the value of this cannot be overstated.

# Open source

Publish and be damned, the saying goes, but the truth of open sourcing is very different. Creating a basic GitHub or BitBucket account costs nothing more than a few few minutes of the usual password creation and, once that's done, you can create a new public repository, to which you can upload any code (or anything else, for that matter) which, for any reason at all, is worth sharing.

Alone the discipline instilled by the knowledge that someone else will be able to read your code is worth the effort of preparing at least some small code fragments for publication to your repository.

And then what can happen? One of the following, most likely:

- Nobody ever reads it, but you cleaned it up and put it up there, so the time was well spent. Somebody might read it next week.

- Someone reads it, thinks it's great and downloads it for use in their own projects. Fantastic, you've made a contribution to the wider community and somebody somewhere thinks you're the bee's knees.

- Someone recommends your repo or links to it from somewhere and suddenly a thousand people have read it.

- Someone reads it and has a suggestion for improvement. This might be the best possible scenario; you are learning something, your contribution to the community is improving in quality, and you have made a valuable contact online.

- Someone reads it, shrugs her shoulders, and moves on. You have lost nothing, you'll never even know, and all the above scenarios are still open.

You have nothing to lose. Don't be overly modest, either. However small your contribution, someone somewhere will be interested in it, you'll have broken the ice, and started to attain a level of comfort with open sourcing and repo's in general. Nothing is up there forever (it's not Facebook), you can delete repo's later if you have something better, or if you feel they're no longer relevant.

If job interviews are part of your master plan, you'll be glad to have built up even a small collection of published code.

# Real-world encounters

You may be fortunate enough to be surrounded by like-minded souls who will happily talk programming with you for hours and days on end, or you may be the only person you know who cares what a compiler is and what it's good for, but, either way, it's a good idea to meet other developers face to face, offline, in a real physical room somewhere. However valuable online exchanges may be, there is no substitute for the wandering, unfocussed, random course that live conversation with a live human being takes. If one stops to consider how many life-changing events come about through some casual mention of something in passing, it becomes clear that it really is worth the effort of leaving the desk and arranging real meet-ups.

## Meet-Ups

If you haven't done so already, check out the Meet-Ups website for a developer group somewhere in your vicinity:

```
http://www.meetup.com
```

If you can't find one (you can guess what's coming now, I imagine), start one! Even if there are just two of you, most of the benefits of live encounters with someone who shares at least some of your programming interests don't depend on how many of you there are. Don't necessarily restrict yourself to iOS or Apple gadget groups, there may be cross-platform groups that are at least as valuable to you as a developer, in terms of knowledge shared, networking, and the shear motivational benefits of a shared passion for software development.

## DevCons

One step up from Meet Ups are the developers' conferences, great and small, that are taking place all over the world, all of the time. Visiting WWDC in San Francisco may exceed your budget (it certainly exceeds mine), but a search of the web will reveal the nearest dev conferences to wherever you live.

The potential benefits of meeting so many other developers in a short (typically one or two days) intensive burst of presentations, talks, discussion groups, and ad hoc conversation over a pizza are too many to list, and too valuable to pass up. Most of them are not expensive, though you may have to factor in accommodation and travel costs.

And if you can't find one near you (you knew this was coming), arrange one! For all you know, you may find yourself in a hitherto unimagined statistical cluster of developers that would like nothing more than a day or two of exchanging ideas, and phone numbers, and GitHub URL's, over a coffee and a pizza. It's always pizza, don't ask why. Somewhere near you is a local community center that will be happy to put a room at your disposal for an affordable fee and you get to add *DevCon* Organizer to your CV and LinkedIn profile. You'll be amazed at the difference that makes.

# Summary

In working through this book, you have learned to plan an app from both a conceptual and technical perspective, taking into account the needs of the user, the desired functionality of the app, and the limitations of what is undeniably a challenging platform, due to its size and memory restrictions. You have then used that plan as a road map to identify and implement the classes and structures that you need to create efficient and robust code that is easy to understand and therefore maintain, and which is easy to reuse in other projects.

This chapter has provided you with some guidance as to which direction to take from this point onwards and which resources are of value to most developers. We have mentioned a few ways in which you can improve your users' initial experience with your apps and add custom navigation, should it be considered appropriate.

We have looked at some of the tools you are likely to need as well as a few tricks that will save you some time. Further, we have briefly touched on topics with which every developer needs at least some level of familiarity, including program design and programming paradigms, version control, and analytics.

We have also discussed the importance of staying abreast of the changes in and around the watchOS platform. Although it is impossible to predict exactly in which directions mobile device use will expand, it is probably safe to say that we are at the beginning of a trend towards an increasing number of ever smaller and more specialized devices that will be part of a larger ecosystem, meaning that the skills you have in iOS and now watchOS development are likely to provide you with an excellent basis from which to expand into whatever developments in mobile programming the future holds.

# One last word from the author

We are, I think we can say, all at the beginning of our watchOS development careers. And it's also the case that both Swift and the Apple Watch itself are very young contenders for the attentions of the world's next generation of app developers.

You, as much as anyone, have the potential to shape the new genre of mobile and wearable devices, a genre that is still finding its feet in the glare of media hype, commercial speculation, and, one hopes, some genuine pioneering spirit among users and developers alike.

As watchOS moves forward and matures, alongside its sibling, iOS, and its distant cousin, tvOS, we should consider ourselves extremely lucky to be in the game, so early in the story.

However important hard work, discipline, a cool head, and even a little financial investment may be to the success of your coding career, whether professional or purely as an occasional past-time, there is one small, obvious piece of advice that will add wings to everything you have covered in the course of working your way through this book, everything you do to support your app's success, and everything that contributes to your growth in software development.

It's fun, so enjoy yourself.

# Index

**Thank you for buying**
# Building Apple Watch Projects

## About Packt Publishing

Packt, pronounced 'packed', published its first book, *Mastering phpMyAdmin for Effective MySQL Management*, in April 2004, and subsequently continued to specialize in publishing highly focused books on specific technologies and solutions.

Our books and publications share the experiences of your fellow IT professionals in adapting and customizing today's systems, applications, and frameworks. Our solution-based books give you the knowledge and power to customize the software and technologies you're using to get the job done. Packt books are more specific and less general than the IT books you have seen in the past. Our unique business model allows us to bring you more focused information, giving you more of what you need to know, and less of what you don't.

Packt is a modern yet unique publishing company that focuses on producing quality, cutting-edge books for communities of developers, administrators, and newbies alike. For more information, please visit our website at www.packtpub.com.

## Writing for Packt

We welcome all inquiries from people who are interested in authoring. Book proposals should be sent to author@packtpub.com. If your book idea is still at an early stage and you would like to discuss it first before writing a formal book proposal, then please contact us; one of our commissioning editors will get in touch with you.

We're not just looking for published authors; if you have strong technical skills but no writing experience, our experienced editors can help you develop a writing career, or simply get some additional reward for your expertise.

## Instant Apple iBooks How-to

ISBN: 978-1-84969-402-5        Paperback: 52 pages

Learn how to read and write books for iBookstore, by fully utilizing Apple iBooks and iBooks Author

1. Practical examples with clear and easy explanations.

2. From a seasoned author, learn the tips and tricks to create a travel guide.

3. This guide also covers the must knows of troubleshooting common errors.

## Apple Motion 5 Cookbook

ISBN: 978-1-84969-380-6        Paperback: 416 pages

Over 110 recipes to build simple and complex motion graphics in the blink of an eye

1. With several images for recipes to aid better understanding.

2. The author has directed films with various celebrities such as Nelly Furtado, Lady Gaga, and Richard Branson.

3. Contains several exercises for beginners and seasoned developers.

Please check **www.PacktPub.com** for information on our titles

## Mastering Apple Aperture

ISBN: 978-1-84969-356-1 Paperback: 264 pages

Master the art of enhancing, organizing, exporting, and printing your photos using Apple Aperture

1. The apt guide for learning advance concepts stepwise.

2. The only book to offer you a plethora of knowledge to not only advance your existing skills but also troubleshoot common errors.

3. A seasoned author, who has worked for various films and shows.

## Moodle for Mobile Learning [Video]

ISBN: 978-1-78216-912-3 Durations: 02:13 hours

Enable learning anywhere with this innovative, and clear video tutorial on Moodle

1. Practical video tutorials with easy to follow examples.

2. From a seasoned author, learn to make learning experience fun for your students by using social media tools.

3. This tutorial will help you revolutionize your teaching skills.

Please check **www.PacktPub.com** for information on our titles

www.ingramcontent.com/pod-product-compliance
Lightning Source LLC
Chambersburg PA
CBHW060517060326
40690CB00017B/3310